ON BEING

Peter Atkins is the author of over 70 books, including *Galileo's Finger: The Ten Great Ideas of Science, Four Laws That Drive the Universe*, and the world-renowned textbook *Physical Chemistry*. A Fellow of Lincoln College, University of Oxford, he has been a visiting professor in France, Israel, New Zealand, and China, and continues to lecture widely throughout the world.

ON BEING

A scientist's exploration of the
great questions of existence

PETER ATKINS

OXFORD
UNIVERSITY PRESS

OXFORD
UNIVERSITY PRESS

Great Clarendon Street, Oxford, OX2 6DP,
United Kingdom

Oxford University Press is a department of the University of Oxford.
It furthers the University's objective of excellence in research, scholarship,
and education by publishing worldwide. Oxford is a registered trade mark of
Oxford University Press in the UK and in certain other countries

British Library Cataloguing in Publication Data

Data available

Library of Congress Cataloging in Publication Data

Data available

ISBN 978-0-19-960336-7 (hbk.)
ISBN 978-0-19-966054-4 (pbk.)

Printed in Great Britain
on acid-free paper by
Clays Ltd, St Ives plc

Links to third party websites are provided by Oxford in good faith and
for information only. Oxford disclaims any responsibility for the materials
contained in any third party website referenced in this work.

CONTENTS

PROLOGUE

The scientific method can shed light on every and any concept, even those that have troubled humans since the earliest stirrings of consciousness and continue to do so still today. It can elucidate love, hope, and charity. It can elucidate those great inspirations to human achievement, the seven deadly sins of pride, envy, anger, greed, sloth, gluttony, and lust. In this book, though, my canvas is more restricted: I consider merely the great questions of being, questions that for millennia have been the inspiration of myth, and explore what science can say to illuminate them and dispel their mystery without diminishing their grandeur or reducing our wonder. I consider matters such as beginnings of universes and selves, and the ends of both.

In preparation for our journey in which we shall nose around among the myths that a collaboration of ignorance and deep concern have jointly inspired, I would like to establish in broad terms my vision of the nature and limitations, if any, of the scientific method. I suspect that few would disagree that science is competent when it comes to the fabrication of novel stuff and novel applications of stuff in general. That, I believe, is not an issue to delay us. Nor shall I linger on the argument about whether these novel stuffs, including better medicines, better and more abundant foods, better fabrics, better modes of communication and transport, better modes of entertainment, and so on, weighed against the social costs, including better ways of killing, injuring our environment, and accidentally or intentionally maiming, add overall to the sum of human happiness. I focus instead on the ability of the scientific method to illuminate matters of great human concern and drive out ignorance while retaining wonder.

The scientific method emerged surprisingly late in human history. Humanity took several thousand years to stumble on to the very simple and to us now somewhat obvious mode of discovery that forms its core: to make observations and then to compare notes. Of course, there are sophistications of both components, for science is not simply a convivial nature walk. The observations are made on domestications of the wild phenomena: they are made on cats rather than tigers (I write metaphorically). Observations cannot be left to blow in the wind of uncontrolled intrusion: they must be focussed and isolated. Nor can observations be merely the recitation of prejudices and preconceptions: they must be *experiments*. The comparison of notes cannot be left to idle chatter: detailed scrutiny and assessment by expert kin is the order of the day. Sometimes, through idleness of the reviewer or deception by the reviewed, error, unintended or malign, escapes the light of intense inspection— but never for long, for scrutiny is endless. Indeed, the more outlandish the claim, the more revolutionary the thought, then the more intense is the pressure that is brought to bear on its assessment.

Then there is the extraordinary marriage of observation and mathematics. Mankind, and for some perhaps sociological reason it has been mostly man kind, has developed a sinewy language of the utmost rigour and austerity that has proved to be the perfect tool for teasing out objectively the consequences of an imaginative qualitative leap or of adding spine to a whim so that it can stand up to the harshness of quantitative comparison of prediction with observation. I hasten to add that mathematics does not appear explicitly in this book, but it does lie as a hidden deep foundation beneath it.

Not all science, however, gallops forward on the back of the grand alliance of experiment and mathematics. Darwin did not formulate his theory of natural selection as a mathematical device, yet it has extraordinary power. His ideas have spawned aspects of the theory that have been capable of being rendered mathematical, and have thereby greatly extended its power. But at its core, natural selection is not a mathematical theory and yet does not lack power. Indeed, it is arguable that the theory of natural selection is one of the most powerful theories ever proposed, for from a simple acorn of an idea, a great forest of consequences follows.

In short, science has as a central principle the publicly shared, controlled observation that we call experiment, and where appropriate is guided, enhanced, and propagated by the austerely logical rigour of mathematics.

So much for the scientific method itself. To what can it be applied and what are its limitations? I consider that there is nothing that it cannot illuminate. Because the scientific method has not yet encountered a barrier, except the one asserted to exist by those fearful of its illumination, my optimism leads me to suppose that the reach of its beam is boundless and in particular that it can replace (or even conceivably confirm) the myths that surround all the great questions of being.

I am aware that extrapolation from present success is not a convincing argument. I am also aware that those, perhaps subconsciously, fearful of science's advance, and also some of our rightly cautious guardians, the more pessimistic of our reflective philosophers of science, will claim that science selects targets that have a soft underbelly where its methods can penetrate effectively and that it skirts topics where either also subconsciously or in fact knowingly it believes its tools to be blunt. In short, some hold that it is a David that seeks out a Goliath with a limp.

But why should anything remain obscure? Some argue that a phenomenon might lie outside the kingdom of the physical and over its border in the neighbouring presumed but physically inaccessible kingdom of the spirit. To those of us who believe, despite the seductive songs of sentimental Sirens, that there is no such place, this argument carries little weight. We accept that there is a great deal of subjective inclination to believe that the physical is not all there is, but as we see no objective evidence for the non-physical, but appreciate that there is a great and forceful sentimental longing for it, we cannot in all intellectual honesty accept its existence as plausible. Sentiment may be an overwhelming mental driving force, but longing is not itself an adequate proof of the existence of what is longed for. We might long to win the lottery, but that does not make it more likely that we shall. Those who promote the spirit might claim to know in their hearts that there is more to the world than the physical, but hearts are unreliable organs of knowledge. That there are billions of people who, given the chance, would vote in favour of there being more to the world than matter and radiation is also not an argument, for reliable knowledge is not secured by majority vote.

Another component of the view that some knowledge might be forever inaccessible by the scientific method is that certain matters are private, subjective, and internal and therefore not open to the rough-and-tumble outdoor brutishness of science with its emphasis on the objective and the public. But science can penetrate into people's heads and into their minds. The mode of inspection is neuroscience and its foot-soldier psychology. Both may still be emerging, and not yet fully reliable sciences, but they are ways to penetrate through the protective carapace of the skull to reveal aspects of the brain's beliefs and sentiments and why they dwell there.

A third component may be that some do not want certain mysteries to be analysed objectively and dispassionately: they cherish their subjective privacy and do not want the mysteries dissected, and by dissection damaged, perhaps diminished, and even destroyed. I cannot accept that ring-fencing certain delicacies in a poetic garden is a valid argument against science's ability to elucidate even the most personal.

If absolutely and unreservedly everything is an aspect of the physical, material world,[1] then I do not see how it can be closed to scientific investigation. 'Scientific investigation' probably sounds clinical and inhuman in this context, something to be abhorred when applied to the tenderness of certain human concerns, but I simply mean open to investigation with the expectation of publicly accessible comprehension, just like any other physical phenomenon.

There is one more aspect of my attitude that it would be inappropriate not to confront in a book dealing with deep concerns: that science is a flowing river of ideas, growing in grandeur and competence as tributaries bringing new ideas join its flow, with occasionally a whole new whirlpool of understanding spinning and overturning what had been presumed to be secure understanding. Thus in the physical sciences Aristotle gave way to Galileo, Galileo to Newton, Newton to Einstein, and Einstein to who knows. Much of scientific understanding, it is claimed, is may-fly ephemeral, awaiting further elaboration or even replacement, so how can I justifiably claim that science has power to illuminate the great questions once and for all?

I do and I don't. Where my account is a review of observation, as in the organic processes accompanying birth and death,

---

[1] Perhaps to the annoyance of philosophers, I shall use the adjectives 'physical' and 'material' interchangeably; I shall further sin by not distinguishing naturalism (broadly, the view that everything there is springs from the natural world and involves no supernatural intervention) from materialism (broadly, the view that all there is can be ascribed to the properties of matter). I adopt the view that the whole of all there is can be accounted for in terms of matter and its interactions.

then there is little force in the view that those observations will be overthrown. Of course they will be elaborated, but the general details and broad features of what I describe are objective, eternal observables, not transient theories. Where, however, my account is a review of theoretical understanding at the edge of physics, then I fully accept that the account is likely to be changed out of all recognition as our understanding of physical reality and cosmology is developed and refined. But my review in these cases will make it clear that our current theories are way-points on the road to presumed complete understanding, showing how far we have come from myth, not concealing that we might have far to go, yet hoping not to quench the sense of optimism that the journey will ultimately be triumphant. Lying between these extremes of confidence and speculative, extrapolated, and currently unfulfilled optimism, are comprehensions of an intermediate kind, such as my account of the origin of species. These intermediate comprehensions are almost certainly correct at their core but are currently undergoing elaboration—with elaboration not to be interpreted as overthrow but enrichment.

In short, I stand by my claim that the scientific method is the only means of discovering the nature of reality, and although its current views are open to revision, the approach, making observations and comparing notes, will forever survive as the only way of acquiring reliable knowledge.

In each of the following chapters I seek to demonstrate the understanding of certain matters of the greatest concern to human beings that has been achieved by application of the scientific method. Humanity, in my view, should take pride in its achievement, that it has unravelled so much and has contributed so many answers to so many of our truly great questions. I hope that even if you are a sentimentally delicate plant unswayed by the storm of intellectual impact that science must be accepted to have had and to be continuing to have and must surely be presumed to be capable of going on having to have, you will not extinguish the spark of pride in what collective humanity can achieve in its quest to understand the workings of the world. Even if you consider that humanity's ability to quest is limited to the physical (as I do, but perhaps for a different reason, believing it to be all there is), surely you can take pride from its journey from utter bewilderment, through first stirrings, through the Renaissance, through the Enlightenment, to our current and far from complete and still evolving level of comprehension.

The following pages are all about shedding myths, acquiring understanding, yet retaining, even enhancing, wonder. I am aware that there are many who consider 'spiritual' and 'material' to be as oil and water. I hope, however, that you will come away acknowledging that it is possible to take a near-spiritual joy from a solely material perception of the world. I hope that you will also take pride in the majesty of the human ability, working collectively in space and time, to emerge from the chrysalis of myth and travel towards true comprehension.

But, just a minute! Before we launch into our journey, there are some pleasant house-keeping duties to perform. This little book went through various transformations, each one guided by the care, insight, and commitment of my editor, Latha Menon. Its final form owes much to her. I would also like to thank my wife, Jean, for her helpful comments during each of its morphs, and Guy Nobes of Marlborough College too. Both contributed really helpfully. There are also lurking in that great cloud of publishing anonymity the reviewers, who made a variety of helpful suggestions as the chapters evolved; I am very grateful to them.

1

BEGINNING

The first great question of being is one that has probably entertained us all at one time or another: where did it, the universe, all come from? How did it begin? That Creation myths abound in a multitude of cultures testifies to the centrality of this question, the biggest question of all. All creation myths struggle to account for the presence of something where previously there was nothing. In some there is a kind of cosmic copulation between the potential parents, with the Earth and its inhabitants a progeny, sometimes rather curiously, perhaps with lice in mind, emerging from armpits. In others, there is a great pregnant cosmic egg, which cracked open, one half forming the vault of heaven and the other in due course becoming the Earth. Sometimes our universe seems to be the result of vented spleen, as in the Polynesian myth in which Ta'aroa, bored with being cooped up in his primordial egg, burst it asunder and in a series of self-defeating petulant adolescent episodes used his own spine to form the mountain ranges, his fingernails for the scales of fish, his intestines for lobsters, and, anger unasuaged, his blood for the redness of the sky. Creation myths of the Abrahamic religions confront the same problem, but do so in what to our ears is a calmer, more mature, more abstract manner, largely by letting God's will conjure order from chaos in some inscrutable way and leaving it at that. Abstraction is taken to its limit in the Hindu *Rigveda* and the *Chandogya Upanishad*, when being was achieved by the negation of non-being; but that is perhaps not a wholly satisfying explanation to every Western ear, coming as it does to within an ant's fingerwidth of being a cop-out.

There are, in fact, three related profound questions to address in the context of creation. One is the mechanism of the coming into being of the universe: what actually happened at the beginning? Another is whether there is any meaning to the question of what preceded the universe and had, in some sense, the potential to become a universe. Here we are confronted by the linguistically and conceptually engaging question of whether absolutely nothing can have potency to become something. The third is whether an agent was needed to trigger the process of cosmogenesis, the process of turning that nothing into what is to all appearances something, or can nothing turn itself into something on its own? All three questions sound as though they might fall within the range of science to answer. A fourth question, *why* there is a universe, is rather different, but still apparently very interesting; I shall set it aside for the time being but return to it later.

One view is that all three questions are immediately and simply solved by positing the existence of a creating agent, which for sake of brevity and to avoid beating around the bush we can call God: God made the universe, in an inscrutable manner. He was outside space and time, able on the wings of a whim to conjure anything from nothing. He was the Prime Instigator, acting for some impenetrable reason, and the uncaused first cause. I shall proceed on the assumption that although such remarks seem more concrete than the extreme abstraction of the *Rigveda*, they are not altogether satisfying. They might, of course, be correct. I shall refer to answers that invoke God as 'religious', for that is what they are, but I shall try to keep God as vanilla as possible, leaving His different flavours of worship to your own selection.

Has science anything better to say? Those who wish to protect a distinct role for religion claim that science, dealing as it does with what already exists and having no experience of what does not yet exist, has no dominion in these questions. They would claim that science is intrinsically incapable of confronting them, with perhaps a grudging acceptance that it might have something useful to say about the newly born universe, when a very primitive version of it was already under way, well after nothing had become something and with the laws of nature already in place. If that view is correct—the view that such important questions are beyond the reach of the scientific method—then the only recourse would seem to be either to accept that the human brain is incapable of identifying the process of spontaneous cosmogenesis or of accepting the existence and incomprehensible activity of a creator.

To assess that position, therefore, I need to explore whether it is conceivable that science is not incompetent in the face of these extraordinarily interesting questions and that some kind of agent-free answer to them will in due course be forthcoming. I adopt the view that the driving force of science is an alliance of curiosity and optimism, the urge to answer questions and the belief that it will answer the questions it investigates; science does not set out on its quest with the underlying presumption that understanding will not be achieved.

First, it is undeniable and perhaps relevant that science has made more progress with elucidating the early moments, if not the inception, of the universe in the past 300 years than religion has made in the last 3000. A more pungent view is that mythical accounts of the origin of the universe are actually admissions of ignorance dressed seductively in engaging allegories: they are stylish new clothes with no Emperor within. Although resonances between these sometimes delightful myths and the formulations of current scientific accounts can be devised, that is perhaps because they are intrinsically vague and, like a rubber sheet, can be stretched to cover more sinewy discussions. We should be cautious, I believe, of the claims of mythmakers that their allegories are prescient of scientific elucidations, for that apparent prescience is arguably a reflection of the elasticity of allegory.

Second, although science is currently seemingly stumped by the details of cosmogenesis, it is important to distinguish 'seemingly stumped' from the actual progress of cautious advance. As a hare might suppose that a tortoise is stationary, so it is essential to recognize the cautious tortoise advance so characteristic of science. Science is an alliance of sparkling imagination and

4

extreme caution. Very rarely do scientists leap to a revolutionary explanation: more commonly, bridgeheads are built and consolidated before edging forward into unconquered territory. The bridgeheads are built on the consensus of theories tested against experiment, and whimsical leaps across a chasm of ignorance rarely survive. Even those great paradigmatic shifts of the twentieth century, relativity and quantum theory, were built on the bridgeheads of classical physics by noting subtle deviations from its predictions. Scientists are revolutionaries, but they are conservative revolutionaries, moving cautiously into and then conquering the unknown.

As a result of their intrinsic caution, almost every scientist is wisely unwilling to express a view about the events accompanying the inception of the universe. Quite honestly, they haven't a clue. Their current task is to edge carefully backwards in time, establishing and confirming insights and attitudes, reaching a consensus, testing ideas against observation, and expecting, perhaps, to arrive at the year dot at an unknown time in the future, being prepared, on the way, to have the entire framework of their, and hence our, understanding overturned by a shift of understanding, and possibly a whole succession of shifts, more cataclysmic than that introduced by relativity or quantum theory.

The quest for year dot—the pushing back by scientists of the frontier of understanding to within tiny fractions of seconds following the inception of the universe, the complex and still largely speculative events accompanying the Big Bang 13.7 billion years ago—has been described in many popular accounts, and will not be recounted here. Suffice it to say that the intricate processes that accompanied the beginning are known, with diminishing confidence and consensus, back to about the so-called 'Planck time' of 10^{-34} s after the beginning.[1] At that point, we know that our current theories fail and that the presumed continuity of space and time, and even their distinction, is lost. In short, we do not yet have under our command the physics necessary to penetrate into the Planck time and go nose-to-nose with the instant of creation. Certain clues to the way forward, however, have been identified, and theoretical physicists are broadly aware of the direction in which their noses need to point. It is certainly going to be a tough journey. It will take a long time to cover that last 100 trillion trillion trillionth of a second, and unimaginable shifts of understanding—so-called 'paradigm shifts'—are likely to be involved. But there is no indication yet that it will be necessary to invoke a creative agent to complete our mosaic of understanding.

That having been said—the 'that' being the admission that currently science can see no better than misty hints of the outline of a coastline of the actual origin, and even that hint might be a mirage, as we mention in a moment—it is still perhaps worthwhile to speculate about the future direction of science's approach to its description of the origin of things. The difference between this kind of speculative—some would say fatuous—exercise and its religious analogues is that the veracity or otherwise of the scientific speculation will be publically accessible whereas that of the religious variety has to be taken on trust. That trust, the religious might argue, is well founded as it is trust in God's word as revealed in Holy Scripture. Moreover, they might continue, we know that biblical accounts are mythical, but those

[1] 10^{-34} s is 1/10 000 000 000 000 000 000 000 000 000 000 000 of a second.

6

myths enwrap deep truths that might one day be revealed and should not be taken at face value by we of the puny mind. Be that as it may, there is in my view a human duty to press on and seek to extend the envelope of human comprehension.

One reason why the misty outline of a possible beginning might be a mirage is that there might not have been a beginning, or at least a local beginning. Certain theories of the early universe suggest that an existing universe, ours, for instance, can give birth to a daughter universe. The implication is then that this universe might have had a mother, and that mother a mother, and other mothers endlessly in the backward abysm of time, and that what we regard as our beginning is not the true beginning of all that is. Thus, our Big Bang is just a local triviality, not a truly cosmic beginning.

In fact, that enlargement of our vision and diminution of our significance is possibly a colossal underestimate of the problem of identifying the beginning. Time might lose its significance on a grandiose cosmic scale and the concept of a 'beginning' be meaningless. It might be the case that any universe can bud into an infinite number of universes, that the current number of universes is already infinite, but increasing, and possibly increasing infinitely rapidly at an infinitely accelerating rate, and has been accelerating infinitely rapidly for eternity, so that our Big Bang is an infinitesimal event on a grandly hypercosmic stage. Although science might seem arrogant in arrogating to itself true understanding, what it discovers is often the foundation of true humility.

It might turn out to be the case that the budding of an existing universe into daughters is much easier to explain than the origin of an initial universe, what I shall call the *Ur-universe*, for at least when a universe exists there are physical laws that govern its behaviour: if we could identify those laws within our universe, then we might find that they entailed its budding into daughters. But even if that can be achieved, there is still the troublesome prospect of identifying the begetting of the Ur-universe. Is God perhaps the begetter of the Ur-universe with His handiwork now irretrievably buried in the myriad descendants that have sprouted since? Without the unreliable assurance of faith, no one knows.

I am forced to wonder if there could still be a beginning even if this breathtaking scenario of infinitely accelerating births of myriad universes were true. There is a way out, but it is wildly speculative, has no scientific foundation, and therefore should certainly not be taken seriously. The only reason for describing it is to emphasize the optimism of science: even when confronted by colossal difficulty heaped on fierce deterrence, there remains the belief that a resolution open to human understanding will eventually emerge.

Just for fun, here is my vacuous foundation-free musing. Let's think of a universe as being made up of dots. For simplicity, I shall consider a very tiny micro-universe of four dots in a line:.... In this universe I can say that dot 2 is next to dots 1 and 3, and so on. The identification of neighbours in a dust of points is technically called imposing a 'metric' and I shall use that term from now on. Thus, our toy universe consists of the four dots with a certain metric. Now I take the next step: the same four dots can have a different metric in which dot 2 is a neighbour of dot 4, and so on, and I can envisage yet another universe in which dot 1 is a neighbour of dot 4, and so on. Thus, the same four dots,...., can constitute several different toy universes.

To take the leap to an actual universe, we can think of it as being made up of an infinite number of dots, with dots 1 and 2, for instance, regarded as neighbours, and an alternative universe built from the same dots but in which dot 1 is somewhere inside a carbon nucleus in a carrot on an Earth-like planet and dot 2 somewhere inside an iron nucleus on an asteroid orbiting a Vega-like star in a distant galaxy. Thus, a multitude of universes can be built from the same dust of points, each one having a characteristic but different metric. In this vision, each of us spans every possible universe: only in this universe am I compactly me; in all the other multitude of universes am I everywhere; in many I am parts of you if you-like entities happen to inhabit them. And because these points constitute spacetime, not just space, I am of all that was and all that will be. In this scenario, the ceaseless budding of daughters is nothing but the flickering into existence of new metrics where points related by one metric enter into a new relationship with others.

I once expressed this wholly speculative, foundationless vision elsewhere,[2] where it was naturally and quite properly met with silence. The thought, though, did at least have one creative consequence, for the American writer John Updike was obviously

---

[2] *The Creation*, by Peter Atkins, W.H. Freeman & Co. (1983).

tickled by it and ran with it much more thrillingly and amus-
ingly in the closing pages of *Roger's Version*,[3] where

> 'So,' says Kriegman. 'Imagine nothing, a total vacuum. But wait!
> There's something in it! Points, potential geometry. A kind of
> dust of structureless points. Or, if that's too woolly for you try
> "a Borel set of points not yet assembled into a manifold of any
> particular dimensionality." Think of this dust as swirling; since
> there's no dimension yet, no nearness or farness, it's not exactly
> swirling as you and I know swirling but, anyway some of them
> blow into straight lines then vanish [...] until bingo! Space-time.
> [...] Out of nothing. Out of nothing and brute geometry, laws
> that can't be otherwise, nobody handed them to Moses, nobody
> had to. Once you've got that seed, that itty-bitty mustard seed, *ka
> boom*! Big Bang is right around the corner.'

One important aspect of this wholly speculative vision is that
the plenitude of apparent creations of an infinitely budding universe is reduced to a single event. The act of creation, from which
our infinitude of generations of universes have eternally sprung
and are currently springing could in fact have sprung from a
single act of creation, the generation of the dust of points that
constitute the basis of each metric. Each individual subsequent
creation then consists of the emergence of a metric. Somehow, or
so it seems to me, the adoption of a metric for a pre-existing set of
points—the flickering of the notion of a neighbour over the surface of an arena of points with each identification of neighbours a
different universe—is a more abstract and therefore less onerous
act than the creation of a fully fledged physical universe.

[3] *Roger's Version*, by John Updike, Knopf (1986).

No doubt all that is nonsense: good stuff for building the character of an earnest and slightly potty fictional scientist. I certainly do not want to give the impression that any scientist thinks that the scenario I have sketched is even remotely supported by any evidence or even theory. I have to stress that all I have sought to show is that *it is possible to think constructively about even the most apparently overwhelming problems* and thereby undermine the view that our inception *must* have been an act of God.

One problem hasn't gone away: where did those points come from? A scientist concerned to see how far rational thought can take him or her has to admit that if at any stage an agent *must* be invoked to account for what there is, then science will have to concede the existence of what we have agreed to call a God. I could express that differently: through the exercise of its godless procedures, ironically science might find that it has demonstrated the existence of God. Because I don't want to concede defeat to the religious and surrender my optimism to faith, and certainly not so early in the book, I need to explore whether science is for ever likely to be stumped by this particular question.

The task before science in this connection will be to show how something can come from nothing without intervention. No one has the slightest idea whether that can happen and, if so, how it can come about. Some consider that the question of the incipience of the universe (I will stop saying Ur-universe, unless I want to emphasize distinction, but that might be what I mean) will be explained by showing that it is a consequence of the laws of physics, such as it being a 'quantum fluctuation of the vacuum'. That, however, is inadequate as an explanation, for absolutely nothing lacks laws just as it lacks everything else, and even a total vacuum is far richer in properties than truly nothing. There are no laws in a universe that does not yet exist, for laws come into existence as the behaviour they summarize emerges with the emerging universe. Tracing our origin to a 'quantum fluctuation of the vacuum' is perhaps appropriate for a daughter universe as distinct from an Ur-universe, for an existing universe does possess a richly propertied vacuum, but it is inadequate for accounting for an Ur-universe, for Nothing has no properties, and thus does not undergo quantum fluctuations.

The unfolding of absolutely nothing—what out of reverence for the absence of anything, including empty space, we are calling Nothing—into something is a problem of the profoundest diffi-culty and currently far beyond the reach of science. It is, however, a target at which science must aim even though to some, even to scientists of a pessimistic or perhaps just realistic bent, it would seem to be for ever out of science's reach. However, because my intention is to show that everything, including Nothing, is within science's reach, and that science provides the prospect of under-standing even the most stupendous of phenomena, I have to travel the optimistic road and, with prejudices flying, try to show that there is hope for a scientific elucidation of creation from nothing.

In fact, I have tried once before to sketch a scenario.[4] My aim now, as it was then, is not to present a theory of creation of something

[4] *Creation Revisited*, by Peter Atkins, W.H. Freeman & Co. (1992).

from nothing but merely to demonstrate that it is possible to *think* about such a process. The actual scientific description of the creation, if and when it becomes available, will be unrecognizably different from what I am about to describe (it would be a hoot, of course, if it were the same!). That is not the point. My aim, I stress again, is to show that human thought need not be stumped by stupendous questions and that it is not inconceivable that science can elucidate the deepest questions of all. That is, I wish to show that there is *the prospect* of an explanation far more deeply satisfying, to me at least, than the allegories suggested by myths.

First, it is important to realize that there probably isn't anything here anyway. I know it seems as though we are part of and surrounded by a material universe, and I certainly would not wish to give the impression that what we perceive is merely a dream or a Berkeleyian solipsism. Of course we are part of and surrounded by things; but at a deep level there is nothing. I shall now try to resolve this paradox, for once it is resolved the notion of creation *ex nihilo*—creation of something from absolutely nothing—is greatly simplified. That is, I shall attempt to show, without I hope unduly embracing metaphysical claptrap, that what I have called the substrate of existence is nothing at all.

The total electrical charge of the universe is zero, but there are positively charged and negatively charged entities within it. We know that the total charge is zero, for otherwise the enormous strength of the interaction between unbalanced charges would have blasted it apart as soon as it had formed. For charges to exist and for the overall charge to be zero, there must be an equal number of positive and negative charges. Presumably before the creation, when there was Nothing, there was no charge; so the coming into being of the universe was accompanied by the separation of 'no charge' into opposites. Charge was not created at the creation: electrical Nothing separated into equal and opposite charges. This 'electrical creation' event was not the manufacture of electric charge, it was the separation of opposites. At the creation, nothing did indeed come from Nothing, but the original Nothing was turned into a much more interesting and potent current nothing when some kind of event split Nothing into electrical opposites. If it was God who provoked the creation and decided to endow the world with electrical charge, then He did not have to make charge: all He had to do was sunder electrical Nothing into opposites. In other words His electrical bounty to us was nothing at all.

The universe does not rotate. In more technical terms, its overall 'angular momentum' is zero. However, locally entities do rotate: locally there is angular momentum, but an angular momentum here (a spinning wheel) is cancelled by an opposite angular momentum (an oppositely spinning wheel) there. When we pedal away on our bicycle, we generate angular momentum as its wheels begin to rotate; but the Earth spins a little faster (if we travel West) or a little slower (if we travel East) and the total angular momentum of the universe remains at its original judiciously endowed value: zero. That is, angular momentum was not created at the creation: then Nothing was separated into equal and opposite rotations. This 'angular momentum creation' event was not the manufacture of a certain endowment of angular momentum; it was the separation of opposites. As in the electrical component of creation, nothing did indeed come from Nothing, but the original Nothing was turned into a much more interesting and potent current nothing when some kind of event split it into rotational opposites. If it was God who provoked the Creation and decided to endow the world with angular momentum, then He did not have to put His shoulder to a great celestial wheel and generate rotation: all He had to do was sunder rotational Nothing into opposites. In other words His net rotational bounty to us was nothing at all.

There is, as we are all aware, a great deal of energy in the universe: galaxies of burning stars, spinning and orbiting planets, and flying balls. Here, at least, the creating agent appears to have had a job to do and has endowed us with, once again, a judiciously chosen and apparently unchangeable bounty. Certainly energy can be converted from one form to another, as when a falling weight accelerates and potential energy (the energy due to position) is converted into kinetic energy (the energy due to motion). But experiments seem to show that the total energy of the universe is fixed and never changes. It is fixed at the abundance that the agent, if any, determined to be exactly appropriate and safe for mankind to have at its disposal and is unchangeable by mankind's subsequent meddling.

That leaves open the question of the magnitude of the initial energy endowment. That can be assessed by totting up the amount currently present, for it will be the same as there has ever been. First there is all the whirling and turning of all the galaxies, the stars, and their planets, and all the little puny events that the inhabitants of these planets consider to be their histories. That all adds up to a colossal amount. But there is an overwhelming more to come. Because it follows from Einstein's theory of special relativity that mass is a measure of energy trapped as a localized entity (the famous equation $E = mc^2$ relating the mass of a region to the energy localized there), to the huge sum we have computed already, we have to add yet another contribution, the energy due to the mass of all the galaxies, their stars, their planets, their moons, their mice, and their men.

But there is another contribution to the total energy that so far has been ignored: the energy due to the gravitational attraction between all components of the creation. Because gravitation attracts, it lowers the energy of two entities that interact, and in particular it lowers the energy of the stars within a galaxy, which clump together under its influence, and lowers the energy of the galaxies themselves because they attract each other over nearly infinite space. This gravitational contribution to the total energy lowers the sum we had previously computed, and there are plausible reasons to suppose that it chips away at that original colossal sum and reduces it—yes—to zero.

No one knows whether the total of all the contributions to the total energy of the universe is in fact exactly zero, but the near cancellation of the positive contributions by the negative (gravitational) contribution is highly suggestive. Indeed, we now know that there is a kind of matter, 'dark matter', with wholly unknown properties except its ability to interact gravitationally with common matter, and which, when included in the sum, brings complete cancellation much closer. Dark matter is highly confusing and its existence particularly humiliating, for although it seems that it is far more abundant than common matter, we know nothing about it except a little about its distribution and have been unable to detect it except by its inferred effect on the formation, structure, and properties of galaxies.

The bottom line, prejudiced with a dash of speculation, is that the initial endowment of energy at the creation was exactly zero, and the total energy has remained fixed at that value for all time. Thus, God's munificence was in fact, it seems, parsimonious on a cosmic scale. He provided us with nothing; nothing electrical, nothing rotational, nothing energetical, and presumably nothing of any other kind. What we see around us is in fact nothing, but Nothing that has been separated into opposites to give, thereby, the appearance of something.

I have taken you through these considerations because it is easy to be overwhelmed by the thought of what had to happen at the creation. The separation of Nothing into opposites still needs explanation, but it seem to me that such a process, though fearsomely difficult to explain, is less overwhelmingly fearsome than the process of positive, specific, munificent creation. The latter raises the question about, for instance, where all our energy comes from; the former diminishes the task of explanation because it reveals that no energy had to be created. Science winkles out the true questions and while leaving the sense of awe intact allows progress with the search for answers.

I cannot resist calling on John Updike again, who was also tickled by this vision of the universe as being engagingly reorganized Nothing:[5]

---

[5] *Roger's Version*, by John Updike, Knopf, (1986).

'This is all metaphor.'

'What isn't?' Kreigman says. [...] 'Think of one and minus one. Together they add up to zero, nothing, *nada, niente*, right? Picture them together, then picture them *separating*, peeling apart.' He hands Dale his drink and demonstrates separating with his thick hairy hands palm to palm, then gliding upwards and apart. 'Get it?' He makes two fists at the level of his shoulders. 'Now you have something, you have *two* somethings, where once you had nothing.'

Kriegman doesn't capture the thought quite accurately, but he is groping towards understanding using reason rather than anecdote. Even he can see that in due course, to understand everything, scientists will be left to study the richness of absolutely nothing at all.

I promised to return to the question of *why* there is a universe. What is its purpose? Something so big, complex, and all-embracing some hold, must be there for a reason.

One answer favoured by the religious is that God decided to make a universe for some inscrutable reason and that it ill becomes us of the puny minds to try to work out what that reason was, or being taken outside time, still perhaps is and will go on being. We could hazard a few guesses, but as they would be inspired by human logic and temperament, and be analogues of the types of decision we might take for doing this or that on Earth, there is little reason to suppose that they would represent anything at all that God might go on having had in mind. Also, it might be considered presumptuous even to think that a human notion of purpose is applicable to so mighty an entity as a God. Put succinctly: we are out of our depth.

Another tack is to try to sniff out evidence of purpose in the universe. Here, science can help and can perhaps lay the question to rest. What does it reveal? Sniff as it might, it finds not the slightest hint of purpose in any events that it has examined. It has put its nose to the ground at the lamppost that represents physics, and found no hint. It has done the same at the lamppost that represents chemistry: not a sign. Biology's lamppost did get more attention than the others because some hold that evolution, which we look at in the next chapter, is driven in a purposeful direction as organisms strive towards perfection (us), but as we shall see, even frantic sniffing there will come up blank.

I need to justify some of these remarks in more detail before turning to the question of whether our local experience on Earth is relevant on a cosmic scale to a universe. To do so, I need to introduce you to that great liberator of the human spirit, the Second Law of Thermodynamics, my favourite law. The centrality of this law to our discussion can be appreciated once we know that it explains why anything happens at all.

In broad terms, the Second Law asserts that things get worse. A bit more specifically, it acknowledges that matter and energy tend to disperse in disorder. Left to itself, matter crumbles and energy spreads. The chaotic motion of molecules of a gas results in them spreading through the container the gas occupies. The vigorous jostling of atoms in a hot lump of metal jostles the atoms in its cooler surroundings, the energy spreads away, and the metal cools. That's all there is to natural change: spreading in disorder. The astonishing thing, though, is that this natural spreading can result in the emergence of exquisite form. If the spreading is captured in an engine, then bricks may be hoisted to build a cathedral. If the spreading occurs in a seed, then molecules may be hoisted to build an orchid. If the spreading occurs in your body, then random electrical and molecular currents in your brain may be organized into an opinion.

The spreading of matter and energy is the root of all change. Wherever change occurs, be it corrosion, corruption, growth, decay, flowering, artistic creation, exquisite creation, understanding, reproduction, cancer, fun, accident, quiet or boisterous enjoyment, travel, or just simple pointless motion it is an outward manifestation of this inner spring, the purposeless spreading of matter and energy in ever greater disorder. Like it or not, purposeless decay into disorder is the spring of all change, even when that change is exquisite or results in seemingly purposeful action.

At its heart, then, all actions, be they physical or cerebral, are purposeless. The process of taking decisions may appear to us to be purposeful because the metaphorical gears in our brains that are driven by the unwinding spring are meshed together in such a complex way, are linked into the chemical repositories of our memories, experiences, and aspirations, and are responsive to stimuli from our environment, that the unwinding appears to have purpose. Indeed, I do not deny the existence of personal purpose, for our brains are such that, in ways that are only slowly being unravelled by neuroscientists, they can identify pathways to pleasure, to fulfilment of various kinds, even to martyrdom if that is their bag. But we should be aware that deep down we, like everything, are driven by purposeless decay: that is why we have to eat.

Now we come up against the crux. Even though at heart all change is driven by purposeless decay, our mental activity ensures that our lives are full of personal purpose. The sense of purposefulness is so great that there is then a natural tendency to extend the notion to cosmic entities. As a result, people have reflected at length on the purpose of the universe, taking the view that if anything is made, then as for human activities there lies purposeful action behind it. This extrapolation from the personal to the cosmic is false.

If, as I believe, the creation of the universe was an agentless act, then it was necessarily purposeless, for there can be no purpose if there is no agent. Even if in due course science has to throw in its towel and, heaven forbid, concede that the universe was indeed created by God, then it remains the fact that there is no trace of purpose on Earth or wherever our telescopes turn. That philosophers and theologians have spent their lives pondering the purpose of the universe is a matter of faint regret, not evidence for its existence.

Why should the universe have a purpose? The question of the purpose of the universe is an invention of human minds, and has no significance, except for the way that it illuminates the psychology of scholarly pursuit and of the pursuing scholars themselves. We should not impose human-inspired attitudes and question on material things. There is considerable grandeur, I think, in the presence of our spectacularly majestic universe just hanging there, wholly without purpose.

2

PROGRESSION

Once we have a universe, we wonder at our position in it and how we organisms have come to be here. How did we humans come to inherit the Earth?

Here has been fertile ground, literally, for mythmakers galore. Just about every creation myth, having dealt in a variety of ways with the substrate of existence, our rocky inorganic planet, elides into the creation of its organic inhabitants. The already dead Mukat of the Cahuilla Indians of California in private conversation with the great shaman Palmitcawut considered that he had revealed the true source of vegetation when he remarked that vines were from his stomach, watermelons from the pupils of his eyes, corn from his teeth, and, somewhat disagreeably to our more tender ears, wheat from his lice eggs and beans from his semen. The Australian High-God Karora seems to have had a problem with bandicoots (loosely, marsupial 'pig-rats'), which emerged in abundance from his navel like rabbits from a conjuror's hat.

He also, seeking company but thereby opening a Pandora's box, gave forth a son, who emerged from his armpit (armpits again) and set about killing the bandicoots like no-one's business. On some nights, Karora's surprisingly fecund armpit was able to produce up to 50 sons at one go, much to the detriment of the now dwindling tribes of bandicoots. The more familiar tale in Genesis is unequivocal:... *God created great sea monsters and every living creature that moves, with which the waters swarm, according to their kinds, and every winged bird according to its kind*...and so on, through crawlies, through creepies, and on to the chief creeper, Man. We are left by Genesis in no doubt that God was our creator but are not brought into the secret about how these acts of creation were achieved, except by the exercise of omnipotent will.

Science, of course, provides a more succinct answer: evolution. That single word raises the hackles of too many for me to assume its universal acceptance and allow me to move on to other things.

Erudite believers of a religious persuasion but liberal disposition accept evolution as a fact and natural selection as its mechanism. While continuing to believe that there is a God, they accommodate evolution by taking the view that God is infinitely subtle and prescient, and that in some sense He enabled evolution as a way of fulfilling His plan for the universe. In His infinite wisdom, they maintain, He established the framework for evolution and then released His primitive but potent creation into autonomic activity and thus allowed it to grow into its full potential and achieve its ultimate aim: arrogantly, but encouraged into the thought by the authority of holy writ, we are content to take that to be us. This view allows sensible believers to accommodate their faith without keeping their fingers crossed or needing to adopt the grotesquely primitive views of the Creationists (with whom I will lump the members of their fifth column, Intelligent Design), who envision God as cosmic manufacturer of every individual type of organism, poring (outside time, of course) over the perfection of a smallpox virus, rendering it exquisitely virulent, designing an eye ten times over, and seemingly cutting corners when it came to his principal creation and allowing defects in design that, even if you and your mother survive the rigours of birth, result in cancer, Hodgkinson's lymphoma, cerebral palsy, and heart disease.

Most believing scientists subscribe to the God-in-the-background view, for it allows them to accept the precepts and apparatus of modern biology and thus to accommodate the irrefutable evidence that evolution has occurred and the compelling theory that natural selection is its mechanism. Their liberal and to my mind largely but not wholly sensible attitude allows them to pursue their craft, some with great secular success. I also suspect that most believing non-scientists of a liberal disposition are prepared to accept this type of compromise position, but I also suspect that many might take the view that He had to give His creation, drifting off-course a little as natural things do, an occasional nudge to achieve His end.

There is, however, a possibility that we cannot ignore: the Creationists—despite their unforgivable manipulation of data, their general failure to respect the achievements of science, their intellectual laziness, and their grotesque distortion of arguments to support their bigotry—might be right. Their view is entirely logically consistent with the premiss that there is an omnipotent God, for omnipotence brings in its train the ability to make, as part of an inscrutable plan or merely on a cosmically capricious whim, worms, lice, HIV, leopards, lilies, bed bugs, and humans, the ability too to leave a trail of false evidence in what purport to be fossils, and the capacity to do all that in a twinkling of a divine eye, or at least in the relative luxury of a working week (or however one interprets 'six days'). Theirs is a compact and succinct account of the origin of everything, which is complete by definition and allows them to lie back and luxuriate in the foam of satisfaction of a job of understanding well done and snipe scornfully through the foam at the misguided who continue to beaver away in laboratories in pursuit of a different kind of understanding. From within their communal bathtub, they can rest in the confidence that they can explain any fact whatsoever with an airy flick of the wrist: they can explain why men have nipples, why cheetah slaughters oryx and blackbird worm, why the penis is used for urination as well as procreation, why *anything* is as it is. There is no apparent reason why they cannot extend their same 'reasoning' to gravitation, without all that fuss about curved spacetime, why they cannot extend it to orogenesis, with God seeing that a mountain range is good, without all that fuss of plate tectonics, and why they cannot extend it to the irrationality of π, without all that fuss and bother of number theory. Luxuriating in their communal tub they can take deep satisfaction from the fact that they have the key to all understanding and that nothing is outside the grasp of their interpretation of 'comprehension'.

25

Unfortunately, their interpretation of comprehension is the abnegation of the intellect. Their approach is to undermine rational thought and to diminish respect for the power of human reasoning. I set aside the charge that they misrepresent facts, in itself an inexcusable evil. Theirs is comprehension of the colour of a leaf by asserting that a supreme agent (which they pussy-foot around calling God, for socio-politico-legal reasons) has designed it to be green rather than to discover its greenness in the evolved composition and concomitant function of its molecules. If their attention were directed at the physical world rather than being confined with almost pathological and suspicious intensity to one aspect of the natural world, then theirs would be a comprehension of gravitation by asserting that the same supreme agent has willed what goes up must come down again. Theirs is comprehension by assertion. Theirs is comprehension that reverts to attitudes typical of the time before the ancient Greeks set humanity on its quest for true understanding. Although the Greeks were mostly wrong in their explanations, at least they *sought* explanation. The intellectual atavism of the Creationists is implicitly to deny the power of human comprehension and to return to days before the first feeble flames of understanding were ignited and the glorious abilities of the unfettered human brain took fire. The Creationists covertly seek a resurrection, but it is the resurrection of the Dark Ages of the intellect.

Creationism is the antithesis of science. Creationism respects written authority (the Bible, despite their cautious disaffirma-tions); science is a ceaseless probing with a view to overturn-ing authority. Creationism implicitly denies the possibility of explanation other than by assertion; science has no truck with assertion. Creationism idolizes complexity by founding its purported explanations in terms of an entity, an agent, a God, that is necessarily more complex than what is being explained; science, though delighted by the glorious complexity of the world, seeks the simplicity that lies beneath it. Creationism denies the power of persuasion of publicly accessible evidence; science relies upon it. Creationism is easy, for it merely asserts; science is extraordinarily difficult, as it seeks covert mechanism.

Creationism is fundamentally dishonest, for as well as having no mechanism of self-control, it distorts the evidence to suit its prejudices; science, though occasionally corrupted by false practice when practitioners are overcome by their own ambition and perhaps thwarted by their own incompetence, has a powerful public procedure of self-policing that sooner or later flushes out malpractice and exposes simple unintended error. Creationism is inerrant: it is devoid of the concept of error and any test of veracity. Creationism is a return to the time before science emerged as a mode of understanding the world; science is ever thrusting forward, wriggling into new modes of understanding, probing to discover wherever it can, seeking to prise open the future. Creationism is deception at every level—it is solid deception, deception through and through; science is the gradual peeling back of veneers that conceal an inner truth.

Creationism is dangerous to society, for it undermines the rational; science contributes positively to society, for it is the apotheosis of the rational. Creationism thwarts the aspirations of humanity; science gives an opportunity for humanity to achieve the aspirations it already has and opens its eyes to new ones. Creationism closes minds; science opens them. Like any thoughtless activity, Creationism is easy. That is part of its danger; for although it is currently a fringe activity, it spreads like a virus into empty, lazy minds elsewhere. Creationism can achieve nothing, nothing at all, except the subversion of true understanding.

Creationism, though not a science, is of interest to science. Just as a dead frog pinned to a dissecting board is a legitimate object of study (and once done in schools when real science, rather than its sociological shadow, the scientific method, was studied), so Creationism can be pinned down and studied. Instead of the frog's entrails, we need to study the psychological and cultural viscera of faith's attack on reason. What drives individuals away from rational investigation? What drives whole groups of individuals to embrace faith, and specifically this peculiar distortion of faith, in place of the intellectual appreciation of the cosmos? How is it that faith can overpower intellect? Maybe it is fear; maybe it is cultural conditioning; maybe it is simple mental laziness; whatever, it is something not particularly admirable in the psyche.

A core foundation of Creationism and Intelligent Design is the assertion that information, specifically the information that the design of an organism represents, cannot emerge spontaneously: information can be created only by an agent. That is nonsense at a variety of levels.

First, we saw towards the end of Chapter 1 (in connection with the Second Law) that although matter and energy tend to disperse in disorder, that dispersal may be used to drive organized structures into being. Information is organized structure, especially when it is embedded in a physical entity, like words on a page and strings of atoms in DNA (of which more in the next chapter). Thus, although every event is accompanied by a net decrease in order of the universe, *locally* order may be generated without the need for an agent.

Moreover, we have to be cautious in interpreting a change in a bit of DNA as the emergence of information. At a molecular level, everything is junk and all changes are random. A protein is built on the basis of the structure of DNA, and in that sense DNA carries information, but the structure of DNA itself has not been constructed with a message in mind. In other words, that a particular strip of molecular junk results in a successfully modified organism leads us *retrospectively* to regard that particular junk as embodying useful information, and we are naturally inclined to wonder how that information arose. In fact, it arose by chance. An analogy might be helpful. A string of 1s and 0s in ASCII computer code might constitute a nonsense word, such as 'thonk'. A change in the digits might occur at random and give rise to any of 'thenk', 'thunk', and 'think'. On seeing 'think' we might cry in triumph 'information!', as it is a viable word. But it is a viability that has arisen at random, and only retrospectively do we regard the message as conveying information.[1] So in organisms as in literature: some modifications of DNA result in viable organisms, others do not. Natural selection takes over once random modifications of molecular junk have delivered

---

[1] I am leaving aside the Creationist's typical riposte that it is extremely unlikely that vast swathes of DNA could arise spontaneously with a huge amount of viable information embedded in them. The simple answer is that DNA did not develop in a little warm pond in one ready-to-wear lump: it grew over time as the reproductive success of successive generations of organisms gradually increased the number of proteins for which it encoded.

a viable message, and if the modified organism is successful in its reproductive capacity it is welded into the biosphere and we think of the DNA as having developed information. The DNA, if it could think like we do, would know that it was just spinning out junk and would snigger at the thought that we had been duped into thinking it to be information. Evolution is not about the purposeful acquisition of complexity: it is about the random generation of successful junk. Instead of thinking of ourselves arrogantly as the apotheosis of creation, it is perhaps more humbling to think of ourselves as currently top junk.

Happily, most of my readers, even those committed to a religious view of the world, will also scorn Creationism as an aberration and I can direct my words at those who are open to rational discussion. To them I now turn. Creationists, who should favour hair shirts, are also welcome to read on.

Deep joy may come from faith, as those who have it commonly say. But deep joy also comes from true understanding, as those who strive for it recognize. There is, as far as we know, no component of the universe more complex than the biosphere, and it is one ingredient of the joy of science that it has winkled out the principal aspects of the mechanism by which that great complexity has emerged.

Natural selection, Darwin's wonderfully simple yet astonishingly pregnant idea, has been written about countless times, and there is no need to enter into details here, for this is a discussion *about* evolution rather than a description *of* evolution. However, there are some thoughts worth clarifying, for there are many potentially muddling aspects of the origin of life and its extraordinary ability to spread over the face of the planet and populate every conceivable niche.

First, it is important to distinguish fact from theory, observation from mechanism, phenomenon from explanation. Evolution is a *fact*; natural selection is a *theory* of how that evolution came about. I think it rather muddling to speak of the 'theory of evolution'. Although natural selection is the currently accepted theory of how evolution occurs, to refer to it as the 'theory of evolution' colours the term evolution to suggest that it, evolution, is a theory whereas it is a fact. This is perhaps a pedantic point,[2] but the issue is of such sensitivity for some people that it is better to be precise.

The observation-based fact of evolution has two strands, two independently rising rivers of empirical evidence that flow together and, where they mingle, unite in support.

The classical evidence for the fact of evolution is the fossil record, a fabric with admittedly many holes and strange rucks and creases, but of such overwhelming coherence that only cosmic conspiracists can doubt its authenticity. Even the rare apparent inconsistencies, such as the remains of an ancestor found in a stratum above the remains of a descendant, can be reconciled by appealing to the ceaseless churning of the Earth and other kinds of adventitious interventions that are not just excuses but are supported by additional evidence. Science is a reticulation of knowledge, a network of interdependent processes and explanations, and its richness and reliability arise from the fact that it

---

[2] So? Science is all about pedantry.

compiles its explanations by drawing from the wells of disparate bodies of knowledge. That an imperfection in the fossil record can be healed by appealing to geology and meteorology is a sign of its strength, not of its ability to contrive deceit.

The second river of information that supports the fact of evolution is more recent: it is now possible to observe the history of the transformations of organisms at a molecular level, at the level of the structure of DNA, not just by digging up and examining, measuring, and comparing petrified bones. That the microscopic sequence of transformations—that expressed in molecular terms—has never been found to conflict with the macroscopic sequence—that expressed in terms of large-scale entities—is powerful evidence that evolution is an indisputable fact.

Indisputable? Well, those who believe in the omnipotence of God will not expect Him to be caught out by failing to match the macroscopic to the molecular. However, even they must concede the *fact* of evolution, unless they think the entire biosphere a plant (in the sense of decoy), although they might have profound reservations about its mechanism.

Then there is the theory of how evolution came about. Here, in principle, there is room for argument and doubt; here is where science behaves like science and allows room for the displacement of one explanation by another. As I remarked in the Prologue, the approach that science adopts in all its fields includes the succession of ever richer views: science is an extraordinary manifestation of the ascent of the intellect. In the formulation of theories, the vitriol of progress can be poured and a once plausible theory seared by new findings or deeper analysis. Here the view that evolution has a teleological component, with an organism striving for some kind of perfection—a Lamarckian giraffe stretching to reach ever higher leaves and thus it and its descendants striving toward the realization of their latent giraffic perfection—can jostle for acceptance with the view that everything is an accident. Here, more in conformity with modern modes of thought, an argument might whirl around whether nurture might influence nature, whether change is always progress, whether natural selection proceeds in punctuated bursts, whether a species might be predisposed to a specific change, or whatever.

A scientist, the arch-descendent of Occam, looks first for the simplest explanation, then builds elaborations only if that explanation's barren, rocky simplicity proves inadequate. Disbelieving scientists regard God as the ultimate antisimplicity, despite the beguiling brevity of His name, for omnipotence surely must spring from complexity. Thus they are reluctant to adopt God as the foundation of any scientific explanation. Far, far simpler for them than the supposition of a God is Darwin's dangerous idea,[3] that evolution is the unintentional outcome of the undirected turmoil of the unconscious competition for resources; in brief, evolution is the result of natural selection. Elaborate organisms—and even a virus is elaborate—have slipped into being as they and their ancestors have stumbled into niches in the ecosystem that have allowed them to proliferate in the face of competition.

[3] I am, of course, aware that I am hinting at Dan Dennett's book of this title (Penguin, 1996).

This harsh view of our emergence, although seemingly god-less, is actually seen by some as evidence of God's extraordinary sophistication. As I have remarked, natural-selection-accepting believers declare that God, in His awesome omnicompetence, endowed an element with the potential for contributing to the evolution of an elephant, and having created the substructure of the world, sat back to allow potential to emerge into the actuality of the biosphere, perhaps by a mechanism that could be identified by science, such as natural selection.

Once again, I have to admit that they might be right, and that a deeply foresightful and patient God conjured an electron out of nothing with Man in mind. It is perhaps easy to be patient if your existence is outside time. However, science does not *need* this elaboration, this jam on the plain bread of its own expla-nation, this velveting of rock, and an equally plausible view is that the invocation of God into evolution is driven by the urge to reconcile a prejudice, or by an emotional need, rather than by the force of evidence.

But suppose we accept that God did lay the foundation of evolution, can we infer from that aspects of what we Earth-bounders refer to as His personality? That God chose the primi-tive barbarity of natural selection to achieve His end, leaving a charnel house of guts through evolutionary history, certainly suggests caution in accepting the conventional reports of His infinite benevolence towards His creation.

Natural selection is all about survival in the presence of lim-ited resources. The immediate problem is *what* it is that is being selected. The facile answer 'the species' immediately opens up a doorway to another question: what is a species? There are many definitions of species, each one recognizing a different aspect of the relationship between organisms, and it should not be too alarming that in an entity so complex as the biosphere a single word is unable to capture the essence of a classification. It is perhaps best that the concept be kept vague for our purposes here lest too rigid a definition freeze thought.[4]

Whatever species are, they evolve. According to the 'modern synthesis', which emerged as genetics illuminated natural history during the early years of the twentieth century, natural selection depends on the heritability of genetic variation, the over-proliferation of offspring, and the suitability of those offspring for their environment. It is a process entirely local in time, being blind to the consequences for a future generation.[5] That in the long term a development might be disastrous is of no consequence to the present: somehow, the descendants must muddle through if and when the conditions change. There is no striving after perfection; there is not necessarily progress, for a felicitous adaptation now might prove to be a burden later. That molecular junk (this is not the general junk to which I referred earlier; this is the numerous dormant strips of DNA that are not transcribed as protein) is carried between our genes might be advantageous, for not only might it stumble into usefulness

---

[4] See my *Galileo's Finger*, Oxford University Press (2003) for an elaboration of this point.

[5] And here, of course, I allude to Richard Dawkins' *The Blind Watchmaker*, Penguin (1990) and shortly to his seminal *The Selfish Gene*, Oxford University Press (1976).

later but it might enable us to revert to a version of our ancestors should a future catastrophe revert to their environment.

But on what entity is natural selection acting: is it on species or individual, or on something else? It is not the species, for each individual strives for its own propagation not that of the hive. It is not the atoms of our constitution, for atoms are on our time-scales immortal and the fact that an atom migrates from eaten to eater is immaterial. If it is neither the species nor the atom, then it is perhaps something in between. Many favour the *gene* as the unit of selection, with the wars of selection being the sub-terfuge of the tussle of genes.

Genes are strips of DNA, as we shall see in more detail in Chapter 3. But even they might not be the actual instruments of selection. Genes convey what turns out, retrospectively as we have seen, to be *information* from one generation to the next, and it is that information that is crucial. Genes, and in particu-lar the molecules of DNA that constitute the physical genes, are merely the carts that are loaded with information. It is the infor-mation that is struggling unconsciously for survival. At its most abstract level, and therefore at its most powerful and general, evolution is the consequence of the wars of information, where one message is pitted against another. Or, if you accept my view that ultimately everything is junk and that information is but the retrospective recognition of viability, then evolution is the consequence of junk wars, with one collection of junk uncon-sciously pitted against another.

The core idea of the theory, the competition for limited resources, and the extreme refinement of this core idea into the sophisticated view that information—as embedded in the physical vehicle of genes, as it happens, and increasingly as actual computer software—is ceaselessly stumbling into channels of proliferation, is almost certainly likely to survive. Not only that, but so compelling is the potency of natural selection that it is virtually certain (and at least in principle a falsifiable prediction) that wherever life is found, in any galaxy, it will have evolved by natural selection.

Even more deeply, and as a buttress to the view that natural selection is cosmically universal, we can regard it as a portrayal of that world-view I sketched at the end of Chapter 1, in connection with the Second Law of Thermodynamics. There, we saw that the spring of all change is the chaotic dispersal of matter and energy and that through the linking of events orderly structures are thrown into being like spray on a rapid as the river-world plunges into disorder. Evolution by natural selection can be regarded as an extraordinarily complex manifestation of the interconnectedness of overall decay, the ultimate manifestation of the workings of the Second Law, with lice, mice, and you and me churned into existence as matter and energy unwind.

Natural selection and evolution are also modes of understanding. The influential Ukrainian geneticist Theodosius Dobzhansky (aka Feodosy Grigorevich Dobrzhansky, 1900–1975) said once that nothing in biology makes sense except in the light of evolution. Organisms are intrinsically bizarre conglomerations of peculiar ingredients, a serendipitously accumulated junkyard of the symbiotic. Whole organs have been assembled by what is often termed Nature's elegant simplicity but which in fact is her unconscious skill in cobbling together whatever is to hand. The only way to understand the outcome, the functioning organ or organism, is to trace its history back through the generations, appreciating how a unit of structure, a cell, a molecule, has been requisitioned and maybe adapted for greater utility or utilization of resources, or greater efficiency at evading another potentially utilizing predator. Those who deny evolution and natural selection deny themselves the ability to understand the function and structure of the living.

Natural selection is not without its difficulties for biologists, just as Einstein's theory of gravitation is not without its difficulties for physicists. However, it is no more reasonable to scorn natural selection because it has difficulties than it is to scorn Einstein's theory. All theories are foundations for more sophisticated theories, giants providing shoulders as perches for later giants.

One problem with evolution is how it began. Competition between primitive organisms is all very well, the triumph of one message over another, but how did matter step across the notional bridge from the inorganic to the organic in the first place? How did signal emerge from noise? An unsophisticated religious view is that God must have breathed life into matter either, as the literalists might believe, into a mound of dust to fabricate an Instant Adam or, as the more sophisticated believers might believe, into a molecule, sending it set fair on its lengthy way to becoming Man, atom *en route* to Adam.

I have tried to be honest in my presentation of the successes and current thwartings of science, and here is another issue where I have to admit that science is a bit stuck. It could accept that God levered the inorganic in some inscrutable way across the gap into the organic and leave it at that. That, though, is not in the spirit of the scientific enterprise. It would be much more satisfying—satisfying of curiosity, intellectually satisfying, and possibly spiritually satisfying—if we could find a physical process by which that gap was bridged, presumably without the apparent intervention of an agent.

But what is the litmus test of life? What distinguishes one side of the gap from the other? How do we know that matter has crossed the bridge and that the inorganic has become organic?

Life is easy to identify but remarkably difficult to define. Too tight a definition excludes what looks like life and too loose captures too much. The capacity to self-replicate is a component of the definition, but not without its problems, as a mule is alive but sterile, and computer software can replicate itself, but we do not, in all honesty, think of it as being alive. It might be tempting to ascribe livingness to an entity that has emerged by evolution, but that would exclude the first living entity and any that we might synthesize from scratch in future. Organisms are organized structures; but so is an integrated circuit. Organisms are organized structures built and sustained by the flux of energy through their interiors and its dissipation into the surroundings; but so are the patterns of convection that can arise in heated liquids and, indeed, the atmosphere, to give rise to the weather: think tornado and hurricane. All known organisms are built from compounds of carbon; but if we succeeded in building a replicating, conscious, self-sustaining, energy-dissipating entity from silicon, would we deny that it was alive? Is a virus alive?

On the whole (by which I think I mean vaguely, in general, more often than not) I consider that abstract definitions are

39

more potent than concrete definitions, for they have a greater reach. We saw a glimmer of abstraction in the brief discussion of natural selection, where I remarked that evolution could be envisioned as the outcome of junk wars, struggles for the propagation of information, the gene being an accidental vehicle, a disposable cart packed with precious information, and replication a device for achieving that information's effective immortality in view of the transience of the physical organism. That being so, on the whole I think I favour an abstract definition of life, one that captures its essence without overemphasizing its physical embodiment. It would also be best if the abstract definition of life were compatible with the abstract description of natural selection and evolution, for the two are so entwined. Life, then, may be *the imperfect sustenance of information by a flux of energy*. A living organism is then a device that achieves this sustenance, perhaps in the absence of its own immortality, by handing on the structural complexity that embodies the information to a sequence of generations that are capable of adapting to competition and a changing environment. Structural complexity—complexity right down to the molecular scale—is information, the absence of randomness, signal not noise. Death is the loss of structural complexity, the loss of information as the flux of energy ceases. With death, we become our surroundings; with death, we as signal revert to noise.

On the basis of this abstract definition of life, our litmus changes colour and signals 'life' in the presence of energy-flux-sustained structural complexity that embodies reasonably stable information. Even if this test is accepted, it helps only to specify the problem rather than to solve it. It seems to suggest that a single molecule cannot be alive but that life switches on only in some kind of aggregation of molecules, an aggregation brought about and sustained by a flux of energy. It leaves open, however, the concept of pre-life, a much more tangible concept than the afterlife we consider in Chapter 4. Pre-life may be an individual molecule that does not alone turn the colour of our litmus but possesses an attribute that is essential to life, such as the ability to generate molecules in its own image, irresistibly an echo of God on His sixth day. There is no particular difficulty in imagining processes that bring such a molecule into existence, and chemists are currently achieving the kinds of reactions needed and per-ceiving that they can take place under conditions akin to those in the early Earth with its widespread hotspots, ocean vents, little warm ponds and great oceans, and stimulating solar radiation. It would be wrong for me to claim that the synthesis of these so-called prebiotic compounds has been achieved under the condi-tions thought likely to be present in those early epochs of the planet, or even as some have surmised, in some kind of interstel-lar medium. However, there is general confidence in the scientific community that some kind of chemical process, as distinct from divine breath, led to accumulations of pre-life material.

That was still not life as we have defined it; ours is a higher hill to climb. Life is the outcome of an alliance of molecules. To achieve the dissipation of energy, an aggregation of molecules needs to form, one molecule perhaps to harvest energy from the Sun or from a local hotspot; another molecule to accept that captured energy and respond; another molecule perhaps to be welded to the second in what at first was achieved by simple incorporation but which in a billion years or so would become a process so elaborate that it would be called a dinner party. The harvester molecule might become entangled with the welder molecule, and the process of assimilating the environment continue. The incorporation of one dinner molecule might give one such aggregate an edge on incorporating others. At that point, evolution would be under way.

This is all entirely speculative. It shows, however, that scientists have not run out of ideas about how the prebiotic gap might be bridged. There is difficulty in finding the actual explanation, because we are uncertain about where it actually took place: was it near a deep ocean vent, in a warm pond, on a lump of clay, in space, in a cloud of volcanic dust, somewhere currently not considered? We are also uncertain about the precise identity of the prebiotic molecules that were first formed and then formed alliances. The question is so important that we do not wish to rush in with untenable explanations. Caution is the watchword of science, caution moved forward and spiced by imagination. At this stage, there is no need to throw in the towel and accept that it was the divine finger that poked rock into life.

The entertainer of a religious interpretation of evolution still has a card to play, for it is possible to regard God as the enabler of manufacture, banker rather than chief executive; perhaps just the purveyor of protons with the potential for progressing to people. This view suggests that God might not have had to roll up His celestial sleeves and do the dirty work of manufacturing the biosphere, perhaps leaving a trail of fossil deception to test our faith in His bountiful nature and infinite love. Instead, He provided a kind of cosmic means to achieve whatever inscrutable end He had in His unknowable mind. I suppose it is conceivable (with God—a being so different from anything we know on Earth, the ultimate in elastic—anything is conceivable), that He did not know what end He had in mind Himself but with what seems to us misunderstanding humans unable to appreciate a broader canvas to be His infinite capriciousness He just set the ball rolling, to settle willy-nilly where it would.

The argument against the view of God as business angel is of only one kind, the same argument that applies wherever His agency is invoked: there is no evidence for it. Although it has the attractive fragrance of harmonious compromise between religion and science, those of us who believe that all assertions are unreliable unless founded on publicly accessible evidence should not allow ourselves to be drawn into the spider's web of false settlement. The scientific account of the emergence of the biosphere, incomplete as it currently is, has not needed to draw on any supposition of agency, either as an ever-present busy-bodying shepherd or as a contented mere initiator.

There is not one jot of evidence—and faith is not evidence—that gives the merest hint that divine involvement has ever taken place. In fact, the opposite would seem to be true. Natural selection is a nasty, brutish thing, with organism set against organism. Fangs, beaks, and claws are the order of the day, not brotherly love: the waterbuck is dead that turns the other cheek. With antelope daily torn to pieces by lion, fear ceaselessly stalks the savannah. If you are religious, you should at least pause before you venerate a God that devised, or just allowed, such a gore-steeped way of ensuring progress and the emergence of the Image of Himself.

Like greater understanding coming from giants standing on the shoulders of earlier giants, the information in our genes has grown through the ages with information pitted against information, serendipitous junk waking up to discover that it is information, in an ever-changing arena. If you favour deep understanding without relinquishing wonder, or more positively and strongly, favour doubling wonder through deepening understanding, then bask in the illumination of that extraordinarily potent idea, that all living things have merely stumbled into their brief interlude of life. Not only are we stardust, we are the children of chaos.

3

BIRTH

The emergence of the biosphere is one thing; just as important to you is how you came to be a tiny and temporary but as far as you are concerned for the time being a hugely important part of it. What really went on between your father and your mother when they took their pleasure, or perhaps made a mistake, and generated the individual you know as I?

Myths galore recount to our personal origins, many relating to the conflicts of motherly and sexual love. It is hard not to think of Demeter (alias Ceres) and Aphrodite (Venus) scratching out each other's eyes. There is domestic Demeter, focussed on nurturing and mothering crops and offspring in general, being circled by the opportunistic, hearth-invading, marriage-dissolving, adulterous Aphrodite, married as she was to the lame smith Hephaestus (Vulcan). Even the bumblebee-like chubby, winged, armed dwarf Eros (Cupid), Aphrodite's son by the importunate Ares (Mars), was a bastard. Aphrodite's birth, such as it was, off Paphos in Cyprus, more or less summarizes what was then known about procreation, for as her name indicates (*aphro* meaning foam, *dite* meaning arisen) she emerged from the surging foam of semen squirted out from the titanic genitals of the Titan Ouranos after they had be cut off by his dementedly aggressive son Cronos and flung splashing into the sea. I am minded to remark that it might still be sensible to take appropriate precautions while bathing in the eastern Mediterranean today.

There is no doubt that had these events occurred in modern times, then the whole family of Greek gods and their Roman reflections would have been taken into care. Like them, for those of a less Olympian nature there was much danger in childbirth, and it is hardly surprising that in most cultures mythical beings were conjured up in men's, and probably mostly women's minds, to encourage fertility and to ease the process of birth. Thus, it is easy to identify all manner of gods and mythical encouragements and propitiations related to reproduction in general and childbirth in particular. Even a few hundred years ago, and in parts of the world still, there was utter ignorance about what went on between conception and parturition. It was much easier in such days, of course, to accept reports that a virgin birth had taken place than it would be in our much more knowledgeable and sceptical times. Science has filled in the chasm of ignorance, and we now know in exquisite detail the facts of the near-miracle that replication represents and have ways other than propitiation of gods to facilitate conception and disendanger childbirth.

In this chapter I shall range over the mechanism of the process of reproduction that resulted in you, looking into both its molecular and physiological basis. There are all kinds of complications—molecular, physiological, personal, social, historical, and international—that stem from Nature's stumbling upon sexual reproduction as a happy if faintly messy and certainly dangerous mode of achieving the propagation of species. One question we need to consider is why Nature did not end up with the apparently self-centred contentedness of the asexual mode, with the conceivable avoidance of all kind of social inconvenience as epitomized by Eve, Helen, and Juliet, or if you prefer Adam, Aeneas, and Romeo, and whoever else is currently highlighted for our titillation in the tabloids.

First, let's consider the molecular roots of our presence on Earth. Biology was transformed in 1953 when Francis Crick and James Watson established the structure of deoxyribonucleic acid, DNA. Before then biology had been largely nature walks. Afterwards, it became a part of the physical sciences with all the quantitative and intellectual power that that membership represents. Molecular biology had been born, and its birth illuminated inheritance.

The molecular basis of reproduction and inheritance, as most people now broadly understand, is DNA. This metres-long molecule has been likened to a long, twisted ladder, with each side of the ladder one DNA molecule and the rungs formed by specific links between groups of atoms on each side. This twisted ladder is the famous 'double helix'. If we were to imagine the rungs as being the length of actual rungs on a typical ladder, then the molecule would be a ladder long enough to reach to the Moon.

The groups of atoms on each strand that form the rungs are called 'bases' for technical reasons to do with their chemical nature, and specifically are adenine (denoted A for short), thymine (T), cytosine (C), and guanine (G). An A base is a bit like a key that fits only a T lock, and a C molecule is like a key that fits only a G lock. Therefore, although the bases appear to be present at random along one side of the ladder (as –ATTGCATGGCCA–, for instance), the other side of the ladder has to match keys to locks (in this case, as –TAACGTACCGGT–), and is therefore the 'complementary' version of the first side. When the two strands fall apart, like a ladder being sawn in half through its rungs, a new complementary strand can form on each separated strand, for the specific key-in-lock connections means that the –ATTGCATGGCCA– strand acts as a template for the formation of a new –TAACGTACCGGT– strand and the original –TAACGTACCGGT– strand acts as a template for the formation of a new –ATTGCATGGCCA– strand. Thus in place of one twisted ladder we now have two identical twisted ladders. That replication of a DNA molecule is, at root, reproduction and its consequence, inheritance.

The order of bases along a DNA molecule, one side of the twisted ladder, one strand of the double helix, is not random. A lot of it is junk, but what isn't junk is you and me. A strand of DNA encodes for the construction of proteins, the worker molecules of our hive of cells, with each contiguous three letter group, such as –[ATT][GCA][TGG][CCA]– in our example, being an instruction to join one particular amino acid molecule

48

(the building blocks of proteins) to another, so building up a chain of amino acid molecules in a definite sequence. Some three-letter 'codons' are punctuation, such as [TGA], the code for 'stop'. The resulting proteins govern just about all the chemical reactions in our bodies and provide many of the structural components (such as our fingernails, for instance) and our protective, hygienic, and variably aesthetic packaging (our skin and hair, for instance). That we and our siblings inherit DNA from both our parents is responsible for us resembling them and our siblings, and for our own offspring not being frogs.

The fact that the replication of DNA is not without error is the root of evolution. As I alluded to in Chapter 2, error provides random variability: in some cases that variation manifests as disease when a protein is built to an inappropriate recipe and fails in its function. In other cases the modification of a protein happens to endow an organism with a functional advantage for a particular ecological niche and the species progresses. At a molecular level, evolution and genetic disease are indistinguishable. The trick, as Nature has discovered, is not to introduce too much error at any one time and to edge further rather than bluster into new territory. Bluster is almost certainly disease; cautiously successful edging is evolution.

Our imagined Moon-distance-long DNA molecule is coiled up inside a cell nucleus that, on the same scale, is a sphere about 1 km in diameter. At certain stages of the lifetime of a cell, as I shall explain, the molecule is distributed over a number of little bundles called chromosomes. Human cells contain 23 pairs of chromosomes, one partner of each pair coming from the father and the other from the mother. Two of these chromosomes are rather different from one another: these are the sex-determining X and Y chromosomes. The Y chromosome is a little squitty thing compared with its X partner, but if it is present (to give an XY pair), then the human is male; if it is not present (as in an XX human), then the outcome is a female. Evolution has not yet found it necessary to contrive a third sex based on brawny YY supermen.

Asexual reproduction is reasonably straightforward at this molecular level: the female (an XX) achieves some kind of budding, and the new bud (which is necessarily XX unless a physiological switch is thrown to turn an X into a Y) is necessarily female. Thus, as the XX female buds furiously, she populates the world with XX images of herself. In sexual reproduction, the males and females of the species achieve a reduction of their chromosome count and certain cells contain, in humans, not 23 pairs but 23 half-pairs. These are the 'gametes', the eggs of the female and the sperm of the male. These so-called 'haploid cells' of the female are all X (because she has no Y chromosomes to start with), but the male's haploid cells are either X or Y, because he is an XY. When a sperm successfully fertilizes an egg, the new individual is either XX (from her X and his X) or XY (from her X and his Y), and the child is either female or male, respectively. To achieve YY supermen, pairs of XY men would need to find a means to mate and civil partnerships would take on a new dimension.

With this molecular basis in mind, I now consider Nature's adoption of sex as a reasonably stable strategy for the propagation of life and its eclipsing, for humans and mammals in general, at least, of asexual reproduction. Parthenogenesis (reproduction without fertilization by a male) is common in plants and well known among certain animals; in the latter it is often a strategy adopted in times of stress. In it, the female simply reproduces herself. The strategy proves to be wise, in natural selection's unconscious way of being wise, because the female populates the land with her own image, other females. As females, like market gardens in general, are valuable plots, this strategy helps to ensure the propagation of species with only minor modification. In other words, parthenogenesis is a kind of marking time until the good times, the males, return.

There are various flavours of parthenogenesis. In the process called 'hybridogenesis' there is common mating between male and female, and the chromosomes of the parents are both present in the ensuing generation. However, when the females produce an egg it contains genetic material that is virtually identical to those it received from its mother, with no trace of the chromosomal material of the father. Thus although the father's input contributes to the male line, it merely catalyses the replication of the female line. 'Gynogenesis' is also a form of male-catalysed parthenogenesis. In this case the male is simply the lubricator of replication rather than its begetter. The female buds as in ordinary parthogenesis, and generates necessarily female replicas of herself, but she does so only when stimulated by the presence of sperm provided by a male. None of that male's genetic material is introduced into the next generation. One problem with this mode of reproduction is that one quickly runs out of males. Certain versions of that peculiar hangover of the Mesozoic, the salamander (and particularly members of the family *Ambystomatidae*, the mole salamander), achieve a supply of stimulating males either by the occasional letting down of her defences, allowing the penetration of a sperm and the chance of giving birth to a male, or by mating occasionally with a closely related species.

If a species has achieved perfection and its environment is unchanging, then parthenogenesis might be a stable strategy. But no species has achieved perfection, despite the claim that Man is in God's image, and the environment is for ever changing as new climates replace old and new predators invade and prowl. Sexual reproduction holds open the opportunity for more rapid reaction to change despite the lottery of courtship and mingling of alien genes, and in the long term seems to be a better strategy. That is, it seems to be better to give up hope of propagating one's self perfectly and to proceed by accepting mingling with similar others in a like predicament. Immortality—the near immortality of information if not of its carriers—comes at a price, in this case the loss of identity.

Now I bring these two aspects of reproduction, a heady cocktail of molecules and sex, together. At a molecular level, the task of the reproducer is to duplicate its DNA, the carrier of precious information and much junk. Nature has found it expedient to use two individuals to achieve this version of immortality, one called a male and the other a female. Within each of these individuals a mechanism known as meiosis (from the Greek for 'lessening') takes place, in which the normal 'diploid' cells or 23 *pairs* of chromosomes (the bundled up packages of DNA) turn into 'haploid' cells containing just 23 *individual* chromosomes each. These are the gametes, the sperm of the male and the eggs of the female.

It is rather easier to understand ordinary replication, the formation of an image of a cell either to achieve growth or to accomplish repair, as in the replacement of skin, so I shall deal with that first. Here Nature is rather profligate rather than well designed, for although not all the information in the genetic makeup of an individual is needed to make a single specialized cell, she goes the whole hog and replicates the entire genetic information, including the junk. I would rather have thought that an Intelligent Designer would have limited a cell's replication to its own recipe rather than recreating the entire cook book as well as barrowloads of penny-dreadfuls. Still, shambolic efficacy is as much Nature's way as her much vaunted economy.

I have to warn you that the following paragraphs are rather complex and in them, for the sake of concision, I shall adopt a didactic voice and, a common textbook device for when the going gets rough, a smaller typeface. I include these paragraphs simply because I want to convey the extraordinary detailed understanding that science has achieved in recent years. You are welcome to skip over them if you do not want to be bogged down in these minutiae and can accept without further ado that science has achieved the near-miracle of detailed understanding.

A biological cell (and everyone is a condominium of billions of cells) consists of an outer membrane containing the protein-rich *cytoplasm*, a fluid with the consistency of treacle. Bobbing around in the cytoplasm is the cell nucleus, where the DNA is stored. We also need to know that the cell contains scaffolding called *microtubules* that hold it more or less in shape, and two important little clusters of microtubules, the *centrioles*, that will come into play shortly. Within the nucleus there are various structures, one being *chromatin*, a tangled skein of DNA.

The events that accompany the replication of the cell fall into two episodes. First is the *interphase*. A sequence of three processes takes place during the interphase. The cell prepares for replication by synthesizing a lot of the proteins it needs. Then the DNA in the nucleus is replicated. I have already sketched how this duplication occurs, with the double helix unwinding and each strand acting as a template for the formation of a new double helix. Finally, when the nucleus contains two copies of DNA, the proteins that are needed for cell division are synthesized and the interphase comes to an end.

Next, the *mitotic* phase (from the Greek words *mit*, thread, and *osis*, process) begins. At the start of mitosis, the tangled, distributed chromatin in the nucleus withdraws into little discrete bundles. These bundles are the 23 pairs of chromosomes, half coming from the father and half from the mother. However, because the DNA has already been replicated during the interphase, each individual chromosome is in fact a double helping of almost identical DNA. Technically, we say that each chromosome consists of two sister *chromatids* joined together at their centres by a unit called the *centromere*. Think of a curly X (don't confuse this with the X chromosome), with a button-like bulge, the centromere, where the two chromatids touch.

More or less at the same time, chemical signals sent out from the nucleus instruct the centriole microtubules to duplicate. The resulting two sister centrioles throw out long microtubules towards each other through the cytoplasm and are themselves thereby pushed to opposite ends of the cell. At much the same time, the nuclear membrane breaks up and liberates the chromosomes within. However, it is only an apparent freedom, because the ends of the microtubules sprouting from the centrioles snake around in the fluid matter of the cell, latch on to the centromeres, and pull the chromosomes into a plane across the centre of the now engorged cell. This process takes a few minutes, and at the end of it there are two notional cells, two proto-cells. There is a centriole in each and the chromosomes arranged so that one chromatid of each curly X faces into one proto-cell and its sister chromatid faces into the other proto-cell. The assembly is still a single cell, and there is no discernable nucleus.

In the final act of this process, the sister chromatids are pulled apart by the microtubules that radiate from their respective centrioles, are surrounded by a newly forming nuclear envelope, and unwind to become the tangle of chromatin. The long microtubules that have done their job of winching stuff around decompose, the cell membrane itself develops a pinched waist that tightens until the two newly formed daughter cells are pinched apart, and where initially we had one cell we now have two, each with its full complement of DNA. We have grown.

So much for replication; now for reproduction. As I have remarked, Nature has found it expedient to achieve sexual reproduction where the DNA of two closely similar individuals is mixed rather than just slavishly replicating the DNA of a single individual. She achieved this by the process of meiosis (as already remarked, from the Greek word for lessening), in which in certain regions of the body the normal diploid nuclei are sliced into haploid nuclei which in males end up in sperm, in females in eggs. Congress of the male and female then recreates the normal diploid cells and replication occurs time and time again, gradually giving rise to the trillion cells that constitute an individual.

Meiosis is a bit like mitosis (this is Nature being conservative again: if it ain't broke, use it; if it is broke, use it for something else), but with a number of subtle but crucial differences. There are two stages of cell division, *meiosis I* and *meiosis II*, and a crucial DNA redistribution step called *synapsis* when the genetic information of the individual's parents is mixed.

In meiosis I the diploid cells that are destined to become haploid gametes undergo the normal busy preparative interphase. As before, in further preparation for being pulled apart, at the end of this interphase the chromatin bundles up into chromosomes, each chromosome consisting of curly-X sister chromatids of duplicated DNA. Remember that 23 of these chromosomes, each with a duplicated replica of the cell's original DNA, come from the individual's father and 23 come from the individual's mother. At this stage, we know the precise lineage of the DNA.

To keep the explanation as straightforward as possible, I shall focus on a single chromosome pair in each nucleus. I shall denote this pair FM, with F the DNA provided by the father and M the DNA provided by the mother. Because the DNA of each chromosome has replicated during the interphase, the F chromosome is now a curly-X pair of chromatids, which I shall denote ff, joined at the button-like centromere. Similarly, the M chromosome is also a curly-X pair of duplicated chromatids, which I shall denote mm.

As the microtubules spin out from the dockside they bring matching pairs together. Thus, the M chromosome is brought up to the F chromosome. This pair of chromosomes therefore consists of four chromatids, mm from the father and ff from the mother. The curly-X ff of F and the curly-X mm of M lie in contact at a variety of positions along their lengths, and wanton promiscuity occurs at a molecular level: chunks of DNA are transferred between chromatids at so-called *cross-overs*. The M and F curly-Xs can no longer be regarded as purely mm and purely ff: they are now mixtures. What was once F has become what I shall call F′ and what was once M has become M′. The F′ curly-X consists of two chromatids that I shall denote f_1f_2 and the M′ curly-X consists of two chromatids that I shall denote m_1m_2. Each of the four new chromatids, namely f_1, f_2, m_1, and m_2, consists of DNA that has been derived, more or less at random, from the DNA that originally arrived as f and m. At this point, therefore, the genetic information of the parents has become mingled and the offspring that might in due course result will bear similarities to both parents and not be exact clones.

Meiosis I now continues in much the same way as mitosis, except that when the dockside cranes pull in opposite directions on the F′M′ pair, the curly-X F′ is pulled as a complete unit into one of the daughter cells and the curly-X M′ is pulled as a complete unit into the other daughter cell. As a result, the process ends with two individual haploid cells, each containing chromosomes consisting of connected pairs of chromatids. One cell contains F′ of composition f_1f_2 and the other contains M′ of composition m_1m_2.

At this point meiosis II begins, starting with the two haploid cells formed in meiosis I. This step is like another mitosis, with the result that the pairs of chromatids that form the chromosomes are, this time, actually pulled apart. At the end of the process, which proceeds just like mitosis itself, there are four haploid cells, each containing a chromatid from the chromosome that separated. Thus, F′ is pulled apart into separate cells, one of which contains f_1 and the other f_2, and M′ is likewise pulled apart into separate cells, one of which contains m_1 and the other m_2, all of them of different composition from the parents but making use of chunks of their genetic material. Where we initially had one diploid cell with a nucleus that contained FM, we now have four haploid cells, each of mixed genetic material.

The next problem for Nature to face is how to deliver the haploid gamete of one individual (the sperm) into the haploid gamete

(the egg) of another, to achieve a new diploid cell that can replicate repeatedly and thus grow into a new individual. The process is still a near-miracle, and we should take deep pleasure that
human minds have been able to unravel it. Unlike the forego-
ing, which was an account of a near-miracle at a molecular level,
I can draw back from the molecular detail and describe this
near-miracle in cell-level terms (but there are molecules playing
in the orchestra beneath).

In the male, the gametes turn into sperm. Certain cells in the
testes, the *spermatogonia*, undergo mitosis, and until puberty
sets in, divide into more spermatogonia. At puberty, the
process of cell division produces two types of diploid cell.
One type remains a spermatogonium (to go on producing heirs) and the other becomes a *spermatocyte*. The latter is
diploid, but undergoes meiosis I and II to become haploid
gametes, the *spermatids*. This little fellow now grows a series of
appendages: its sharpened head for penetrating, if it is lucky,
an egg, and a motile tail that can move it through the variety
of fluids it encounters during and after intercourse. A sperm
is arranged very much like an Apollo Moon rocket, with the
genetic material in the capsule at the top, and the equipment
for driving it forward below. A principal difference (apart from
not being driven by rocket motors, an awesome prospect) is
that the fuel for its motion is drawn largely from the sugars in
the fluids it moves through.

In the female things are rather different, for an egg requires
rather more attention than a sperm and the extraordinary near-
miracle within her takes rather longer than in the male. Her
childbearing capacity is determined in the foetus, for there the
diploid cells that in due course will produce eggs are formed and
replicated by the normal processes of mitosis, and at birth she

has nearly half a million of these so-called *primary oocytes* in her primitive ovaries. Only about 500 of these potential people will actually be released to the prospect of fertilization in the course of her lifetime. The process of meiosis—the division into four haploid gametes—begins shortly before birth, but is arrested and remains patiently incomplete for 12–14 years throughout childhood. At puberty the process of meiosis resumes, but with great caution and with an eye on the special need of a fertilized egg for nourishment.

At the onset of puberty, some of the primary oocytes awaken each month and of those activated oocytes one or two undergo meiosis I to form two haploid cells. However, unlike in the male and with an eye on future needs, these two cells are quite different in size. One, the *secondary oocyte*, hogs most of the cytoplasm and the other, the *first polar body*, is a runt. The latter might or might not undergo meiosis II to give two haploid cells, but they have no future and in due course decompose. (Those who consider that every egg is a potential person with a right to life might reflect on the fact that God seems to think otherwise.)

The secondary oocyte is where the future, if any, lies. However, even after all these years it has not gone beyond meiosis I and inside its nucleus lie the pairs of connected chromatids with mingled genetic information arising from the various cross-overs that have occurred. In fact meiosis II (the separation of these chromatids and the formation of a pair of haploid cells) occurs only if a lucky sperm penetrates the cell's defences. If it does, meiosis II immediately occurs—this is the lean economy of meiosis on demand. However, as in the female's version of meiosis I, the cells are quite different in size, with one hogging the cytoplasm and the other, the second polar body, a runt destined for decomposition and oblivion without issue.

Two runts and one fertilized ovum fat with nutrients, two abortions for the sake of one child, have emerged from this lengthy process, with a little stimulation from the male. The nutrient-engorged fertilized ovum needs its food supplies for its seven-day journey down the fallopian tube before it makes wombfall and can start to draw its nutrients from the mother. From now on it is on its way to becoming a person and, in due course, repeating the processes that brought it into being.

Such, in general terms, is the story of how you became. I think there are two points to absorb. One is the extraordinary complexity of the mechanism that Nature has adopted. Some sit back awestruck and overwhelmed, and maintain that such complexity could not have come about by the blind gropings of evolution and must have been designed and engineered by some kind of agent (aka God). Others, the real scientists, sit back awestruck but not overwhelmed, and accept that natural selection has extraordinary potency and through more or less random rivulets of exploration can arrive at wide rivers of wondrous complexity. It would probably be impossible to predict, starting from a unicellular organism, that such explorative rivulets would lead to that complexity, but Nature, through the current existence of that complexity, has shown that it can. Scientists are still puzzled about how this complexity emerged under the impact, presumably, of natural selection, and it remains a problem of evolution. That is not to say that there are not many ideas about how it came about. Just as for the origin of life itself, which is still a real puzzle, evolutionary biologists are not without ideas, but have not yet identified which, if any, is valid.

Another point is that the unravelling of the mechanism, at both its molecular and cellular levels, is a remarkable testament to the power of human collaboration, which shows what can be achieved by meticulous internationally shared public investigation; in a word, by science. We are not merely stardust and the children of chaos: we are the spreaders of light.

4

DEATH

From birth and the cradle, I turn to the fag end of life, death and the grave. Puzzlement, apprehension, and the inability to come to terms with the prospect of one's own annihilation have been powerful inspirations of myths relating to death and few are able to look it straight in the eye and not flinch.

Perhaps nowhere have myths relating to death been so elaborately industrialized and thoroughly entrenched in almost every aspect of life as in ancient Egypt, where they overwhelmed reality, become a national obsession, and dominated the economy. The Egyptians considered man to be made up of six parts, three vaguely material and three seriously spiritual. The three material, but only just, components were *khet* the body, *ren* the name, and *shut* the shadow. The three spiritual, and very definitely so to the point of obscurity, as is perhaps appropriate for matters spiritual, were the *ka*, the *ba*, and the *akh*. The *ka* is perhaps closest to what we think of as the soul, being responsible for the perpetuation of the person into eternity. Stimulated by the observation that bodies did not decompose at a great rate in the aridity of the desert, it was supposed that in the underworld the *ka* went off looking for its previous corpse, and upon identifying it, entered it, so restoring it and its owner to eternal life. Should the *ka*, despite anxiously and diligently mooning around for ages and turning over myriad once-dead corpses, fail to find its owner, then it would result in a second, more final death. Thus, the Egyptians placed a considerable emphasis on the preservation of bodily remains, for should those remains decompose too much, the *ka* would fail to recognize its former embodiment and that would be that.

The heart was held to be the centre of self, intellect, and emotions, a view common in many cultures and which still pervades our language, and was regarded as much more important than the rather curious passive stuffing of the head, the brain, which didn't seem to do very much at all except to keep our heads in shape. So whereas the latter could be discarded, commonly by hooking out through the nose, and the empty skull filled with resin, pitch, or linen, the heart was represented by a scarab that could speak up for its previous owner when Osiris called it to account. The scarab was sometimes signed so that there would be no mistake and was kept close at hand by the expectant and hopeful corpse. Even the common people were in with a chance at immortality; pharaohs were in up to the hilt with their elaborate mummification procedures, and made sure by the construction of elaborate masks that their *ka* would not simply miss them in the bustling crowd of partially decomposed dead.

In our less arid zones, bodies do decompose, and there has been correspondingly less presumption about the physical preservation of bodies for personal resurrections. We all expect to decompose, either in the quick flame of a crematorium or the slower belly of a worm, and an honest and brave approach is to consider what is actually in store for us rather than the hopeful and baseless anticipations of our distant ancestors.

I shall base this account on my own death. That, at least at the time of writing, had not yet occurred. But one day it will and it is of some interest to me to know what will happen to my body, for it is an old friend, we have been together for many years, and I am still rather fond of it despite its various idiosyncracies and mounting imperfections. Although this is my *post mortem* autobiography, you should be able to adapt it to your own forthcoming demise, which I hope is not imminent, for I speak of the bodily decay that inevitably visits us all once we cease to live and our bodies are not immediately consigned to a flame.

To spin the yarn I shall make a variety of assumptions that might or might not come about in each of our cases. On occasion, as the whim takes me, I might consider myself shot, the victim of a traffic accident, or just left for dead in shady woodland. To maximize the potential for overlap with what will actually bring about my death and yours, I shall not go into the medical condition, the accident, the murder, or whatever it is that is actually in store for me and to give a little breathing space to my account I shall assume that my body was left undiscovered for some time and was not cremated. I shall make use of various scenarios so that the ground I cover has a reasonably wide scope.

As I did in the discussion of the valley of birth, I shall guide you through the valley of death in two stages. First, there are the straightforward physical phenomena, such as the changes in body temperature and the nasty sloughing off of skin. Then I shall increase the magnification of my microscope, and see what death means to our molecules. My account is, I hope, an appropriate combination of sensitivity and truth; be assured that I did not take pleasure from writing it, except in so far as it enlightened me, and on rereading, it disturbs me.

A hot inanimate object, a rock, a block of iron, or even a cup of tea, cools according to Newton's law of cooling, that the rate of cooling is proportional to the temperature difference between the object and its surroundings. As a result, an inanimate object cools fast at first and then progressively more slowly as its temperature approaches that of its surroundings. The exact rate of cooling for any object depends on its composition and size; technically, it depends on a property known as its 'heat capacity', with objects of high heat capacity, effectively being hungry for heat, cooling slowly. Water has a high heat capacity, which is one reason why ice forms slowly on lakes in winter and why the oceans are a kind of thermal ballast and help to stabilize the temperature of the planet. In so far as a human body is mostly water, it cools quite slowly to the temperature of its surroundings with the precise rate depending on the extent of thermal contact with them.

A human body, though, is not a simple inanimate object, even when dead. A cadaver is not a cup of tea; it is a complex collection of different tissues that undergo chemical reactions as they decompose and thereby generate heat. *Algor mortis*, the cooling of a body following death, is highly variable, depending as it does on the clothing of the body and the local conditions, and little reliability can be placed on the old rule that the rectal temperature falls by slightly less than 1°C per hour from its typical 'living' value of about 37 °C. In practice, a dead body cools slowly for the first half hour or so, and sometimes longer, and then more rapidly before levelling off again as the body approaches the temperature of its surroundings. Typically, the rate of this last stage of cooling is high when the period for which initial slow cooling occurs is short. In some cases, especially when there is infection, the temperature might actually rise after death.

My rectal temperature might have been quite low at my time of death, especially if I had suffered trauma and was in shock. I would have started to cool rapidly. Just how rapidly would have depended on a variety of factors. First, it depends on body mass, with large bodies cooling more slowly than slight bodies on account of their greater water content (remember the oceans) and obese bodies more slowly than trim on account of the insulating character of fats. I am not particularly obese, so my body is more pond than lake and my post mortem secondary cooling would have been correspondingly quite rapid. From the empirical correlation between that rate and the initial slower rate of cooling, it is perhaps also safe to infer that that slow period did not last for long.

Clothes insulate and retard cooling, especially if the clothes cover the region of the lower trunk. Once again, for the sake of this discussion we can suppose that I was only lightly clad when death befell me. It is reported that the cooling of a naked body is about half as fast again as when fully clothed. We can presume, therefore, that I cooled rapidly both on account of my slightness and my near nakedness. Wet clothes advance the rate of cooling both by conduction and the evaporation of the water. Under certain circumstances, perhaps having been confronted and attacked by thugs, I have no doubt that I would have wet myself and be weeping in fear and agony.

Cadavers cool more rapidly in even lightly moving air. The ambient temperature is an important factor and it could easily have been a quite chilly day when I collapsed on my walk in the wood. Cooling is faster in humid air because its thermal conductivity, its ability to conduct heat, is greater.

The loss of heat by conduction is not an important factor for a living person, for whom radiation, convection, and the evaporation of perspiration are the principal modes of heat loss arising from the typical 100 watts or so at which humans typically operate and is fuelled by the metabolism of foods. 'Insensible heat loss' is the loss of heat by the largely constant rate of evaporation of water from the lining of the mouth, from the lungs, and through the skin and accounts for about 10 per cent of the background generation of heat by that metabolism and bodily activity. Sweating, an active process, is switched on by the hypothalamus when its sensors in the skin and in the interior of the body sense a potentially dangerous rise in temperature. The blood vessels in the skin also dilate and the warm blood circulates more freely there, and heat loss by radiation, conduction, and convection is enhanced. The evaporation of the increased superficial water produced by the sweat glands is very important because even the evaporation of a small amount of water can absorb a lot of heat. Thus, if there were no compensation, the evaporation of 1 litre of water draws over 2000 kilojoules of heat. That is enough to cool 60 kilograms of water from 37 °C to 0 °C and to freeze nearly 10 per cent of it to ice. During hard exercise we might generate as much as a litre of water an hour.

I might have lost a lot of blood in the accident or attack and be in hypovolemic (low blood volume) shock. My peripheral temperature might have already been quite low due to the vasoconstriction that would have consequently occurred, and I would have been unlikely to be losing much heat by sweating, radiation, and convection. However, if I had been quickly transferred to a cool place and my body had been in intimate contact with cool surfaces, losses by conduction could have become considerable. Forensic pathologists need to take the details of body–surface contacts into account when seeking to estimate the time of death from measurements of core temperature.

The moment of death is marked by 'primary muscular flaccidity': the muscles relax and the victim slumps. Hollywood has captured this flaccidity by gunshot thousands of times. If I were shot, then I would likewise slump to the ground and lie there lifeless. Within a few hours, *rigor mortis* sets in and my joints are immobilized. To understand the mechanism of the onset of rigor, we need to examine the structure of muscle ('little mouse') and its mode of action that gives rise to the little mouse-like ripples under my skin as they contract and relax, and delve a little more into life and death from the point of view of those ultimate onlookers, our molecules.

To set the scene we need to know that life, from the professionally dispassionate viewpoint of a scientist, is the avoidance of a certain kind of equilibrium; death is the usually unwilling achievement of that equilibrium. Conception, gestation, and birth build the apparatus for the temporary avoidance of equilibrium; dying is accompanied by that apparatus's final, irreversible loss.

The 'equilibrium' of which I speak is not simply life–work balance; it is a technical term relating to the direction of chemical change. All chemical reactions, and in particular the glorious panoply of chemical reactions that constitute the processes of life, have a tendency to proceed in a specific direction. Those of a Taoist inclination will believe that there is a force that urges these life processes in a particular direction. We scientists know, on the other hand, that the driving force is much more prosaic, well defined, and fully understood: it is the natural tendency of matter and energy to disperse in disorder as expressed by the Second Law of Thermodynamics and as discussed in Chapter 1. The details of how a chemical reaction is driven forward constructively by that unconscious agent chaos, though, need not distract us here.[1] Thus, for this reason, and keeping track of changes in the location of energy as well as the location of atoms, hydrogen and oxygen tend to form water; water has no

[1] *Four Laws that Drive the Universe*, by Peter Atkins, Oxford University Press (2007); also available as *The Laws of Thermodynamics: A Very Short Introduction*, Oxford University Press (2010).

tendency to fall apart into its constituent hydrogen and oxygen. To generate hydrogen and oxygen from water, we have to force the water molecules to decompose, perhaps by passing an electric current generated by a battery. In contrast to water, the natural direction of change of a protein molecule is its decay. To build that protein molecule, certain reactions have to be driven in their unnatural direction and the components, in this case amino acids, forced to link together. This forced linking and the resultant construction of a protein is achieved by coupling the process to other reactions that have a strong tendency to run in their own natural direction.

The fall of a weight provides an analogy. The natural tendency of a weight is to fall to the ground. If we want to raise a weight, we can do so by coupling it with rope and pulley to a heavier already raised weight and allowing the latter weight to fall. To raise that heavy weight into its working position we must allow a yet heavier already raised weight to fall, and so on. That is why we have to eat. The metabolism of food is like the heavy weight falling, the biochemical processes in our body are elaborate versions of the rope and pulley, and the formation of a protein from its components is the light weight rising. The food we ingest, the heavy weight raised, is created by the fall of the heaviest weight of all, the energy released by the Sun, with the elaborate rope and pulley of photosynthesis and its elaboration and harnessing, agriculture.

One very important heavy-weight reaction involves a molecule that plays a role in every cell of our body, adenosine triphosphate (ATP). This middle-sized tadpole-shaped molecule consists of a bulbous head and a tail of three phosphate groups (a cluster of phosphorus and oxygen atoms). The equivalent of the falling weight is the losing of its final phosphate group, leaving the truncated adenosine *di*phosphate molecule (ADP). The coupling of this loss to other chemical reactions occurs widely throughout the cells of every organism and is one of Nature's most common processes. It takes about 30 such processes to build a link between two amino acids in the process of building a protein. The raising of this weight so that it can act in this way is the attachment of a phosphate group back on to an ADP molecule, restoring it to being ATP. That is the process driven by breakfast, lunch, and dinner.

Equilibrium is when all the weights are resting on the ground, with all ATP lying around dead as ADP, like guests at a Mafia shoot-out. On a cosmic scale, equilibrium is when the Sun has gone out and is no longer able to drive lighter weights upwards. On a personal scale, equilibrium is when all our weights have fallen, all our ropes broken, our pulleys scattered. No longer can a protein be built from its components; all the proteins have decomposed. All our elaborate structure has decomposed: gone are the proteins that catalyse our reactions and constitute our structural filaments and packaging, gone are the lipids of the membranes of our cells, including those repositories of consciousness, our neurons; gone are the little molecules that scurry through the body with their vital messages, gone are the carbohydrates that power our movements and the movements confined to the brain, our thoughts. Gone, then, is that most personal thought of all: I am.

My life, like yours, is the temporary avoidance of this state of equilibrium. The elaborate physical and chemical organization of my body that grew following conception and birth is the intricate busy web that contrives to keep raising the weights on high. The weights are not resting on shelves in a state of static equilibrium, for that would have meant that I would have been an unresponsive organism: a statue, not a living being. The responsiveness of me the organism, my being alive, depended on the weights being mobile. To continue, develop, extend, and gloriously mix this metaphor: the weights had to be held high but not shelved; nubile, not old, maids; a beehive in summer, not a potting shed in winter; Oxford Circus on Saturday, not Wall Street on Sunday. That is, the network of chemical reactions in my body needed to be ceaselessly active and sustained by metabolic activity. When the network was impaired, this activity was no longer sustained and I the organism ceased to function: I died.

With ATP in mind, we can begin to understand the onset of my *rigor mortis*. A single muscle is a great bundle of fibres, each fibre a single cell with numerous nuclei formed by the fusion of many embryonic cells and reaching lengths of up to about 30 centimetres. The internal structure of each fibre is also fibrous, being composed of bundles of fibrous molecules. For muscle, think fibre, fibre, fibre, right down to fibrous molecules, all bundled together.

Four kinds of molecule are important for this discussion, namely myosin, actin, tropomysin, and troponin. I shall unfold their roles. To imagine a myosin molecule, think of a rope with bulbous ends, bend the rope to bring the ends together, then twist it many times. These individual molecules then lie together to give a 'thick filament' with a knobbly surface. Lying between the thick filaments are the 'thin filaments'. (Scientists are not always imaginative or arcane in their nomenclature.) A thin filament is composed of four strands of molecules twisted together. To imagine this structure, think of an actin molecule as two lengths of string twisted together; then wrapped around these twisted strings to reinforce them and help hold them together are two wire-like slender tropomysin molecules. Embedded along the length of the thin filament are individual troponin groups of molecules. These molecules act like press studs: they bind to the string-like actin and to the wire-like troponin and help to pin them together. They also bind to calcium ions (electrically charged calcium atoms) if any are present.

At rest, the thin and thick fibres lie between each other, but there are lengthwise gaps. When the muscle contracts, the knobs of the thick filaments effectively walk along the thin filaments, the degree of overlap increases, and the muscle shortens. Whenever I did anything when I was alive, such as when typing this chapter or anything else that involved motion—walking, pointing, speaking, pontificating, breathing, blinking—my thick filaments were walking along my thin filaments, then letting slip again.

Let's think about what went on at a molecular level when I blinked. Once again, the detail I shall now go into is not crucial to the overall argument and the next three paragraphs can be skipped. However, I include it because I want to convey the astonishing depth of understanding that scientists have achieved in their unravelling of the processes that aggregate into the extraordinary phenomenon we call life. You might also appreciate the extent of the rolling back of events that would have to take place to restore a dead body to life.

The signal that initiated this process—the process by which my brain signalled that I should blink, the link between my thought and my action—was the entry of calcium ions into a muscle cell in my eyelid. In the resting state the wire-like tropomysin molecules that spiralled round one of my twisted string-like actin molecules blocked the receptors on the actin and the knobby heads of the thick filament could not attach to them. When calcium ions arrived in response to a signal travelling along a neuron, they attached to receptors on the press-stud troponin molecules distributed along the thin filaments. As a result, the troponins became slightly distorted. This distortion allowed the tropomysin wires to slip a little around the twisted actin string and thereby to expose the binding sites for the myosin knobbly heads, which now could snap into place and become poised to claw their way along the filaments. The snap into place of the knobs dragged a short length of the myosin rope out and away from the rest of the thick filament. These events, triggered by the influx of calcium ions, were driven by ATP. An ATP molecule is associated with each of my myosin knobs, just as they are in you, and as it discarded a phosphate group the molecule released energy and decayed into ADP. The heavy weight had fallen, and in its fall had pulled out the claw-like myosin knob.

At this stage in the contraction of my eyelid muscle, the myosin ropes were bent both where they left the thick filament and where the knobs ended the ropes. When the molecule was bent like this, the ADP molecule that had fuelled the action could escape, leaving the angles of the two bends to become more acute. As those angles changed, the thin filament was ratcheted along and my muscle contracted a little. In its new shape, after it had pushed the filament along, the myosin head became associated

74

with another ATP molecule, the link to the actin was broken, and the head and its newly acquired ATP could sink back towards the thick filament and be ready to start this sequence again. The thin filament could not slip back because not all the heads on the thick filament fall away at the same time, so the actin was pinned into position until the myosin linked again and ratchetted it further forward. This sequence occurred many times very quickly, and my eyelid muscle contracted by about a third of its initial length. The muscle relaxed again when the calcium ions were pumped out of the cell, the troponin molecules reverted to their original state, and the tropomysin threads slid back over the myosin receptors. The Velcro-like connection between the filaments was lost and protein springs drew the muscle back into its resting position.

So much for the abundant movements of life. In death, rigor arises from the step where the attachment of ATP causes the knobs of the thick filament to bend back from the thin filament. If there is no ATP present, the two filaments stay bound to each other and the muscle cannot relax. We are stuck and rigid.

As we have seen, the presence of ATP, the raised heavy weight, depends on biochemical processes that force a phosphate group back on to ADP. In death, those processes cease, the supply of ATP is not replenished, the filaments stay locked, and my eyelid, like the rest of my body, became rigid. It stayed that way until the proteins that constitute my muscles started to decompose; then rigor was lost as 'secondary muscular flaccidity' set in and I became flexible again.

Rigor sets in progressively at a wide range of rates that depend on the identity of the muscle and the state of the individual. Because the muscles do not contract but are effectively locked into the position they had assumed at death, rigor more or less preserves the configuration of the body at the time that it sets in, and only in exceptional circumstances does it contort the body. This is different from the contortions of bodies killed by exposure to heat (think of Pompeii), where high temperatures coagulate the muscle proteins and they do contract.

Although the molecular processes of rigor would have been broadly simultaneous throughout my body, their effects would have appeared first in my small muscles, such as those associated with my eyelids; then it would have crept up my limbs from the small muscles of my hands and feet to the bigger muscles associated with my arms and legs. Rigor can be complete in some people after only about 3 hours, but a few may go for as long as 12 hours before it has set in fully. The rate depends on a variety of factors. It is faster if the dead had starved before death because then the supply of ATP runs out more quickly. It has a more rapid onset after vigorous activity (once again, due to the depletion of ATP) and at higher temperatures; rigor rarely develops in cadavers kept at about 10°C and below.

Secondary muscular flaccidity, the flexibility that arises from the decay of muscle protein, sets in after about 36 hours after death at ordinary temperatures but can be achieved more quickly at higher temperatures. It occurs more quickly when the onset of rigor has been rapid. Because it is cold at night and we can suppose for the purpose of illustration that the place where my body rested undiscovered for a day or so in its woodland setting was probably shady and cool, we should perhaps presume that this state of decomposition had commenced by then and I would have been flexible again.

'Post mortem lividity' is the dark purple discolouration that arises from the pooling of blood in regions close to the skin and is often easily apparent in the earlobes and the tips of the fingers below the nails. Blood normally coagulates when exposed to air but that ability ceases within about an hour of death due to the release of enzymes, fibrolysins, which decompose fibrinogen, the fibrous matter that forms clots. The accumulated blood is typically purple due to the continued detachment of oxygen molecules from haemoglobin and the mingling of veinous and arterial blood. Lividity begins to appear at about half an hour after death, beginning with the formation of red patches that intensify in colour, acquire a purple hue, and coalesce into bigger blotches; it is normally complete within about 10 hours.

The lividity of my undiscovered body would continue to develop. Had my discovery and transfer to the mortuary been later than about 12 hours after my death, the initial pattern of lividity would have become permanent and a new pattern would have been superimposed. My internal organs would also show lividity, had they been inspected. The fact that I was lying down would have resulted in regions of compression where my blood would not accumulate, and as a result there would be patches of contact pallor. The hair of corpses is often said to continue to grow. It doesn't. It just appears to grow as the skin contracts and the hairs simply become more prominent. At that stage, scarred, wounded, blotchy, and if not already bearded then at least with a 5 o'clock shadow, I would be best kept discreetly out of sight.

By then, I had started to smell. Bacteria are ever present and although many might cohabit in the host, as in the colon for a digestive symbiosis, they are false friends in death. Bacteria would have spread from my bowel and started to consume their erstwhile host. The rise in acidity of bodily fluids that accompanies death and the elimination of oxygen from the soft tissues, favours the growth of anaerobic organisms (those that do not depend on oxygen). As a result, their trail of digestion results in gases and fluids that have not undergone oxidation. Thus, the sulfur present in many proteins is released as hydrogen sulfide (H_2S, the smell of rotten eggs) and closely related noxious compounds, and the nitrogen that is a universal component of all proteins is liberated as ammonia (NH_3) and its relatives the amines. These foul smelling compounds have names that speak for themselves, such as cadaverine and putrescine.

The rate of putrefaction is strongly dependent on temperature, and in cool environments takes place only slowly. Despite my wounds, which provided ample invasive routes for bacteria and flies, we can surmise that I decomposed quite slowly in the cool of the glade; but decompose I did.

Sulfur is a neighbour of oxygen in the periodic table of the elements, that map of chemical relationships, the brotherhood of matter, and consequently resembles it in some of its chemical properties, but with subtle variation. Thus, whereas a haemoglobin molecule is bright red when oxygen is attached, it is green when sulfur attaches instead. Once body parts start to decompose, there is plenty of sulfur around (as its compounds), and as oxygen leaves the body, sulfur can take its place. As a result, starting at about 36 hours after my death a green hue would spread over my body radiating from my large intestine, rich as it was in bacteria. As this discoloration spread, the veins close to the surface would have become visible as a purple–brown network, especially on my torso and in my groin. Physical changes take place as the skin turns a dark red-green and 'skin slip' might have occurred, with large sheets of epidermis sliding off at the slightest touch.

After about a week so much gas would have been generated had I lain there undiscovered that my abdomen would have become distended and would have pushed out a putrid, bloody fluid through my nose and mouth, and, mixed with faeces, from my rectum too. This same gas is generated within all the tissues of the body as they decompose, and my whole body would have started to swell, especially where the tissues were loose. At this stage, my face was green-purple, swollen, my eyelids swollen and tightly closed, my cheeks puffed out, and my swollen tongue protruding through my mouth. Soon after, my brain turned almost to liquid and my internal organs became unrecognizable.

If my corpse were left above ground, then it would have quickly become a moving mass of maggots. If buried it might remain in reasonable shape for months but within a year all the soft tissue would have been eaten. All that would have remained would have been my skeleton and teeth.

I am glad that I am at the end of this discomforting chapter. However, it is interesting, I think, to know what changes our physical bodies undergo once the sustaining flux of energy, in the form of food, is interrupted and metabolic processes cease. We need to know that we stardust, we children of chaos, we spreaders of light, are inescapably destined to decay.

5

ENDING

There are two kinds of immortality that we need to consider: one on this side of the grave and one on the other side. Provided the incompetencies and indignities of senescence could be avoided, immortality on this side of the grave, the one that medical advances might conceivably achieve, would be fascinating. After the novelty had worn off, though, perhaps going on for ever would be considered a little tiresome and perceived as somewhat selfish. The belief in immortality hoped for by many members of mankind, though, is quite different from life eternal on Earth, and is supposed to take place on the other side of the grave or the door to the crematorium oven. Who, when the ideas of eternity and immortality were first formulated, would want to be perpetuated on Earth, with its violence, dirt, poverty, stench, and disease? Life then was nasty, brutish, and the shorter the better. No; the eternal life available once through death's door was an altogether more splendid existence in a realm not of this and happily free from every pain from toothache to gangrene and without a hint of either actual or metaphorical grime. The good of heart will have been distilled from the bad of act, like fragrant oil from stinking tar. To our everlasting satisfaction, the bad—be they ruffian or king—will have got their come-uppance and be undergoing delicious and ceaseless torment in Hell. Everything will be very, very nice; it would be a kind of celestial Poundbury without the irksome presence of incompatible neighbours.

As far as relevant myths are concerned we are in the world of 'eschatology', the discourse on Last Things (from the Greek words *eskhatos*, last, and *logia*, discourse). Eschatological matters are of the highest importance to some, for they illuminate the whole point of being. The faithful take the view that matters of the First Importance are illuminated by the discussion of Last Things, for they, the latter, are the consummation of being and the apotheosis of existence; in short, things not to be sneezed at. To the more sceptical, there is the suspicion that nowhere else in speculative discourse has so much endless nonsense been written. The sceptical, I suspect, consider that if normal theological discourse and myths in general are the Himalaya of nonsense, then eschatology is the Martian Olympus Mons, towering miles above petty terrestrial Everests.

So far, I have confined my brief mythical introductory *hors d'oeuvres* to the past, where it is quite comfortable, being so detached by the separation of time, for us to snigger at the naivety of our ancestors' fanciful suppositions, especially when those ancestors inhabited different lands: were foreigners as well as ancients. Now is the time, though, to wheel on to the stage some current myths, myths that are said to be believed by some even today. No longer shall we have the luxury of historical and geographical separation to insulate us from those who hold them. As usual, after reviewing some of them, I shall bring us all back to Earth, or what will remain of it when we sit out approximations to eternity, with an account of what I think actually awaits us, whatever our beliefs.

The sceptical, who typically regard various religious claims as low fruit for rational discussion, consider that in the brimming orchards of eschatology they wade through windfalls. This is where they step on squashy plums, overripe grapes, engorged figs, apples of such luscious ripeness that even the wasps weep with joy; this is where they are knee deep in the end of the world as we know it, the resurrection of the dead, the Last Judgement, Heaven and Hell, and the consummation of God's purpose. Whereas the religious see such matters as of the highest importance and as representing the apotheosis of their faith and their reward in Heaven, the sceptical consider themselves up to their necks in metaphorically overripe passionfruit, lengthy bananas of the brightest yellow, and pears of such enormity that beside them pineapples are as peas. For here, at least according to the views of the more extreme believers in the Christian tradition, believers who are still among us (but, I hasten to add, not any of the more sober believers who occupy the middle of the convoluted highways that we commonly regard as religious roads), are the four donkeys of the apocalypse: Millennium, Tribulation, Armageddon, and Rapture.

The *Millennium* is not the largely arbitrary, calendrical, secular one we enjoyed a few years back but the one we are told about in that handbook of horrors, the Book of Revelation. It will be a period of 1000 years of unsurpassable joy, a period of universal peace that will enable us to see that everything that we have gone or are about to go through—the Crusades, the Black Death, World Wars 1 and 2, even the 2012 Olympic Games—was worthwhile.

Rather more troubling will be the *Tribulation*, a mercifully short period of a mere seven years when the Antichrist comes to power and provokes *Armageddon*, a time of fearful war when hardly a soul will survive. Death, destruction, and despair; all will be furiously and unforgivingly visited upon the Earth.

Happily, there will also be the *Rapture*. How we should crave and long for the Rapture! Provided, that is, we happen to be Born Again. Those who are without faith, and those who have been merely once-born and have not been Born Again, will watch with seething envy as the dead Born Agains emerge triumphantly, smugly, and perhaps a little superciliously, from their graves and are swept up in a mighty column to their future glorious abode in the sky. These are the business-class Born Agains, who will board Heaven first. The apple green of that envy in the bosoms of the Left Behinds will deepen to darkest emerald as the economy-class Born Agains, the Born Agains who are still regrettably alive, jostling along with their carry-ons, form a multitudinous throbbing throng that sweeps up to the welcoming arms of their Maker and Redeemer, who is uncontrollably sobbing, shedding tears of heavenly joy at the final rapturous ascent of His faithful.

The Rapture, those in the know hold, will be sudden, and all the better for it. People will be swept up from their beds, their baths, and their bicycles; the finest yachts will be left bobbing crewless off Antigua, rickshaws will be abandoned in Viet Nam, driverless cars will topple from mountain passes in Peru; passengers on aircraft will be puzzled and understandably concerned about the sudden disappearance of the cabin crew once it has dawned on them that it is not the normal absence of service; they will also start to wonder if anyone is left in the cockpit and hope that the pilot at least is an atheist.

It is in the common nature of myths that there is often a certain degree of disagreement among the knowledgeable about the exact sequence of events that will take place. That is certainly the case for our four donkeys. According to the absolute certainties of *historical premillennialism*, the Millennium will follow Armageddon, and its 1000 years will be literally bliss on Earth. According, however, to the absolute certainties of *amillennialism*, we are already well into the Millennium and even well into the Tribulation, as a quick glance at any daily newspaper will confirm. Those who are not too fond of the thought that they are already caught up in the Tribulation might, therefore, prefer the alternative absolute certainty of *postmillennialism*, where the Tribulation is played down and it is held that Christianity is already spreading like necrobaciliosis over the surface of the Earth, converting Jew, Muslim, Hindu, and Buddhist as it goes, as the ever more brimming churches and increasingly over-burdened pews of England so convincingly attest.

With what to a dispassionate observer strikes one as a reticence formed of being once bitten, twice shy, as soon as He is confident that all the Jews have been safely converted into something less threatening, Jesus comes back to Earth, resurrects dead believers—provided, presumably, they have not lost their faith while they were dead and, perhaps, allowing dead unbelievers to correct their errant ways in the contemplative quiet of the grave—and performs a Last Judgement, weighing our insubstantial souls much as Osiris had to listen to the pleadings of scarabs.

As might also be suspected, there are equally conflicting absolute certainties about the precise timing of the Rapture. The choice here is between the unarguable certainty that the Tribulation will precede the Rapture and the equally unarguable certainty that the opposite is true and that the Rapture will precede the Tribulation. Historically, compromise has usually been far from theological debate, where hands typically have been found to lie itchingly closer to swords, but so important are Last Things that happily one is available here, for it is also held to be absolutely certain by some that the Rapture will take place exactly in the middle of the seven-year Tribulation. Spoiling for a fight, though, equally certain to others, brooking no dispute, is the fact that the Rapture will come close to the end of the Tribulation, with a rump of the latter still to run. Happily for the Born Agains, it is also absolutely certain that for them, but not for the unprescient chumps who have not got themselves Born Again, the Rapture will enrapt true believers just in the nick of time at the start of the Tribulation leaving the erstwhile chumps the chance of quickly getting Born Again and being rapt at the end of the Tribulation.

A sure sign that something of this kind is already up or that its upness is imminent will be the sighting of Christ in the process of His Second Coming. Dates have been advanced with absolute certainty about the best time for making such a sighting, which presumably might resemble the opposite of an eclipse. A common feature of all such past predictions is that none has come true, unless Jesus II has been particularly reticent in declaring Himself and has slunk back to Heaven disgruntled that no one had recognized Him. There have been numerous disappointments of this kind. The year of the Great Disappointment was 1844, when everyone had been on Black Alert for Him and very chipper; the following year they got the potato famine instead. Another certainty was 1914, but if that was Him, then He had chosen a manner that tested the faithful to the limit.

The only chilling thought among all this persiflageous disputation is the possibility that powerful Born Agains, with their fingers close not to swords but nuclear buttons, will conspire to bring about Armageddon and thereby, at the expense of civilization, murderously verify their ludicrous but professedly sincerely held beliefs.

Let's be sensible and let these donkey distractions yield way to a discussion of seemingly more serious issues. The first of the Last Things we shall consider, wearing our serious hat and using our scientific eyes, is the notion of the afterlife. Longing, the sense that surely life can't be this bad, and as I have already alluded to in my discussion of that trapdoor of life, death, namely the inability of people to come to terms with the prospect of their own annihilation, have jointly inspired the deeply held belief, a belief encouraged by scriptural remarks and in so many revered texts of other religions, that there is not only

86

better but best to come. Christian orthodoxy teaches that this life is but a necessary pre-afterlife: it is but a time of trial, a time for the faithful to be filtered from the unfaithful, and the good to be distilled from the bad. The real living, know the faithful, starts once you are dead; accordingly, they long for it, but with the exception of certain martyrs seem to take every precaution to postpone it.

As we have emphasized throughout these chapters, the spine of the scientific method is publicly shared evidence, not the wishful thinking of private longing: public experiment rules, not private sentiment. In the present case, the message from science is starkly simple and hardly needs to be said, but say it I will: there is no evidence for life after death and it is contrary to all our understanding of what constitutes life to suppose that it exists in any sense at all. Immortality is indeed achievable, be it Mozart through music, Newton through physics, and Tutankhamun through tourism, but in every case it is an immortality of the footprint they have left in the course of their life, not immortality of a physical body or spiritual soul.

As usual, I have to acknowledge the normal defence erected by those determined to believe that there is an afterlife, that science has gathered its experiences on this side of the grave and supposes only falsely that its gaze pierces through the iron curtain of death and can comment on the world beyond. How can we be sure, they ask, or—more strongly—how can we have the arrogance to suppose, that knowledge acquired in the laboratories of the living has any relevance to the realm of the dead? Such a view would be powerful if there were any evidence for an afterlife; it loses its potency completely in the absence of any.

I shall identify life, for the sake of this chapter, not with the abstract, sparse definition that I suggested in Chapter 2 (the imperfect sustenance of information by the flux of energy), but with what we perceive to be the core of our own human existence, namely some kind of consciousness, an awareness of self, surroundings, and others. It could be that in the afterlife each of us has as much awareness as a toadstool; but surely there is little point in going to the trouble of having an afterlife if you are unconscious of it. Science fiction writers might surmise, still in the absence of any evidence, about the conglomeration of all the deads' consciousnesses into a global corporate whole, thus forming a sufficiently powerful organ to think of useful ways of passing the time during eternity rather than just hanging there uselessly in a state of bliss, like on an endless aircraft journey in first class. Such a view would be powerful if there were any evidence for it; it loses its potency completely in the absence of any.

It is commonly held that the substrate of a post-life existence, collective or personal, like its actual-life precursor, is consciousness. But we know that consciousness is an outcome of the neuronal actions that constitute the operation of the brain, and with brain gone, so is consciousness too. Dualism, the fantasy that Mind is distinct from its substrate Body as represented by Brain, is just that: fantasy. Shoot away bits of the brain, and consciousness fades: HAL regressed as he/it was disassembled. Nature does her own vile shooting in Alzheimer's disease: bits of the victim's consciousness do not gradually seep over to the Other Side as the physical brain decays and awareness and cognition dim on This Side.

The brain is the centre of production of the sense of self, of the conscious determination to survive, and great acts of creativity, all of which are components of what we term, for ease of discourse, the 'human spirit'. But that is all it is: the 'human spirit' is a portmanteau word for that package of intentions and achievements: there is no substance beyond the verbal packaging. There is absolutely no evidence for the existence of a brain-free miasma-like emanation that corresponds to a self's actual spirit. The faithful's hopeful elaboration of consciousness, the Soul, *ka* writ large, is nothing but a metaphor for the sense of self. It is certainly true that those professional pessimists the philosophers, like the professional optimists the scientists, have made little progress in understanding the nature of subjective experience, with *qualia* and the sense of I; but as a member of the latter I at least am confident that all such aspects of consciousness are rooted in the physical brain and one day will be fully illuminated. There is no evidence whatsoever that the self is imbued with some kind of virtue that can survive the decomposition of the organic, physical body and the signal-processing power of the brain it contains.

I think that to deny this analysis the faithful will argue that the Soul is a concept that transcends our notion of consciousness, typically by using words that have no meaning or by arranging meaningful words into sentences that make no sense: the Soul is a consciousness beyond consciousness, a transcendental hyperconsciousness. They will argue, I suspect, that Soul is a spiritual quality, a quality undetectable by laboratory instrumentation built as it is from Earthly matter intrinsically insensate to the spiritual, like trying to listen to the radio with a seashell. For them, the Soul survives even when matter-based consciousness has expired and given up the ghost. The Soul, they imply, is the dark matter of consciousness, undetectable except by implication. Unfortunately for this view, there is absolutely no evidence, the seductive whisperings of the heart, scripture, tradition, and theological musing aside, that it is true.

Primitive believers, who may range from pagan to pope, believe in the reality of ghosts and view them, even if they do not fully articulate the thought, as denizens of the underworld and hence manifestations of, and therefore evidence for, the afterlife. Once again, ghosts can be dismissed as airy nothings by appeal to *reliable* evidence, of which there is none. Certainly there are frissons, creaks, whistles, plonks, and things that go bang in the night; but these all have the simplest of physical explanations, including breezes that blow and buildings and their fitments that settle in the relaxation that follows the stress of the day.

Then there are the claims of communication with the spirit world. Communication with spirits, the lifeblood of spiritualism, is commonly the outcome of clever probing before the seance and theatrical contrivance during it. All these amusements dissolve into airy nothing when inspected by science or by conjurors who know the tricks of the trade. Without exception, all spiritualists are sharks feeding on the gullibility of the weak, distressed, and hopeful who inhabit the oceans of the world.

And what about those who have travelled to the threshold of the Other Side, have caught a glimpse of the promised land, but have been hooked back by a busy-body surgeon in the nick of time? Near-death experiences are widely documented and the reports are often detailed. Unfortunately (for Afterlifers) they are devoid of supportive content but of considerable fascination to physiologists who see them as a window on the processes that accompany the gradual shutting down of bodily and particularly mental processes.

The afterlife would be a harmless deceit if it were not such a potent weapon of sometimes malicious control. Indeed, for the utterly hopeless, those trapped irredeemably in poverty and disease, and perhaps those on the brink of their own death and desperate for any solace, then it may have a therapeutic role and alleviate hopelessness with the prospect of hope, even though that hope is an illusion. There is little to lose on your deathbed, except your life. There is a lot to lose, though, before you get there.

The second of the Last Things we shall consider is the resurrection of the dead. I dealt at length with death in Chapter 4 and will not warm over the body of that discussion here except to remark on the impossibility of being able to restore a cadaver to working condition, even one quite freshly dead. There is little point in emphasizing the almost infinitely greater impossibility, if impossibilities can be magnified, of assembling the bits and pieces of all past, present, and future Mankind, some scattered to the winds, some digested by maggots, some dispersed, some blasted into fragments, some gone up in smoke, some dissolved in acid, and some gone, literally, to the dogs.

I am aware that a lot of people really do believe that they will be reconstituted as bodily beings. Some are reasonably enough concerned and perplexed about the state in which they will find themselves when marshalled at the Last Trump. Will their liver still be cirrhotic? Will their hernia be reversed? What about their infarcted heart and sclerotic arteries? Will they still limp? Indeed, will they get back the foot, leg, finger, hand, or arm they had lost a decade before? Will they still be mute, deaf, or blind? Will nonviable human foetuses, the ultimate in meekness and thus presumably the ultimate in deserving, come back too with a better chance or be just as unviable as before? What about the aborted? What about your acne or your paunch? Might you still be bald? Varicose veins? Would you by the grace of God be able to choose your best moment? Perhaps there will be a one-look-fits-all unisex perfection, a blend of Helen of Troy and Michelangelo's David or an aesthetic dumbing down in the general direction of the Elephant Man? Perhaps everyone will come back as a kind of politically correct cosmic average. Even some ostensibly clever people have wondered seriously about such questions, for if you take your bodily resurrection seriously, then such questions are real.

I believe that the only real question raised by this particular Last Thing is how people can take it seriously. It raises a *psychological* question: how can it be that seemingly sensible and educated folk cannot come to terms with the fact that death is extinction? In particular, never mind the others, who cannot accept that *their* death is *their* extinction? 'Faith' would be their response; but a psychologist would seek to peel back the skin of this remark and examine the underlying flesh of explanation. Those who cannot countenance extinction as their fate, cannot imagine the world going on more or less the same without them.

The penultimate of Last Things is the end of the world as we know it. Science can contribute helpfully and precisely to this discussion, for it can provide a reasonably reliable post-history of our impending future.

It is an arguable stance that pagan Sun worshippers were unknowingly right in directing their respect to the Sun: almost every organic activity on Earth is driven by the energy released in vast quantities by the Sun as it burns away in the sky. Should the Sun go out, then life on Earth would quickly wither and die. The weather would be thrown into turmoil, all vegetation would yellow and die, the food chain would be extinguished, and the concatenation of consumption that allows the captured energy of the Sun to trickle down from grass through sheep to Man would cease. That would truly be the end of days. If anything is worthy of worship, then it is our 4.52 billion year old Sun, the bringer of light and the bringer of life.

The Sun will go out. But the going out will not be simply an extinguishing of its light, for the events that will accompany it will be slow but cataclysmic nevertheless. To understand its death we need to understand its birth from a great cloud of gas, mostly molecular hydrogen and a tiny smattering of contamination of other elements. Nearly 5 billion years ago that great cloud started to fall in upon itself as gravity drew the molecules together, with the contraction becoming noticeable after 100 000 years and the accumulation recognizably solar after a million years. The condensation of 2 million trillion trillion kilograms of gas, the equivalent of 330 000 Earths, under the influence of that awesomely weak force, gravity, did not occur in a rush but began, at least, as a leisurely but, as we now know, portentous infalling.

As the molecules fell together and collided ever more vigorously, so the cloud heated up and the molecules were smashed apart into atoms. Some of the collisions were so vigorous that some nuclear fusion could begin, and the sprinkling of deuterium atoms (heavy hydrogen, a neutron stuck to a proton rather than a proton alone) in the collapsing cloud began to fuse together and release energy. That release of energy added to the rise in temperature due to gravitational collapse. There was another consequence: the turmoil in the cloud was now so great that electrons were knocked out of atoms. Now, instead of electrically neutral atoms, there was a plasma (a gas of electrically charged particles) of nuclei and electrons. A plasma absorbs light far more effectively than do neutral atoms—we can see deep into space, because we are looking not through plasma but through the countless neutral atoms that populate it—and suddenly the cloud became opaque. A watcher would have seen the sphere begin to glow, and where there had been obscurity there would have come illumination. The Sun had come cautiously, tentatively alight.

Still the process is leisurely, and it took another 7 or 8 million years before the temperature of the pregnant and now dense cloud became high enough for the real business of nuclear fusion, the melding together of nuclei, to begin in earnest. After 34 million years since the cloud first began to collapse, the nuclear fire had fully taken hold and hydrogen nucleus was smashing into hydrogen nucleus and producing helium (in essence, two fused hydrogen nuclei). The actual nuclear processes that were taking place, and still take place, are more complicated than that, but that is their outcome.

Four and a half billion years ago, the incandescent cloud, fuelled by the energy released by nuclear fusion, was recognizably the Sun, surrounded then as it still is today by a few specks of old cold matter that failed to condense into the central sphere and which we call the planets. Some of the matter on one of those specks drifted into intelligence and now understands how the Sun was born. Just as astonishingly, the same matter knows what will happen as the Sun continues to burn its fuel.

The radius of the Sun is currently 700 000 kilometres, about 100 Earth radii. To imagine the size: think of 100 penny coins, each one representing the Earth, laid out in a row: that, on the same scale, is the diameter of the Sun. The middle 20 coins mark out the diameter of the core, where half the mass of the Sun is concentrated, where the temperature is about 16 million degrees, the density is 12 times that of lead, and where 99 per cent of the luminosity of the Sun is generated. Here in the core is the engine of life, the region where the main business of nuclear fusion it taking place, with hydrogen being converted into helium, the energy to fuel our 400 trillion trillion watt Sun corresponding to a loss of mass of 5 million tonnes a second as nuclei fuse together.[1] The energy released

---

[1] Einstein's iconic equation, $E = mc^2$, written as $m = E/c^2$, shows that the energy, E, in a region can be measured by determining the mass, m, of the region, c being the speed of light. So, if energy is released, the mass decreases. Mass is not *converted* into energy: it is a *measure* of the quantity of energy.

in that process is radiated through the medium forming a spherical shell 40 coins thick (so, together with the 20-coin core, making up the 100 that represent the overall diameter of the Sun). But the journey of the radiation is highly convoluted, for it scatters hither and thither, and even travelling at the speed of light takes a million years to reach the outer edge of the region.

Once there, the energy is brought to the outermost surface of the Sun by turbulent convection through a spherical shell of the 20 outermost coins (20 for the core, 30 on each side of the core for the radiative inner region, 10 on each side of that for the convective region). Convection is the bodily motion of matter, and photographs of the Sun's surface show the tumult there as hot matter rises and cools—relatively cool but still at thousands of degrees—and sinks again. The temperature here is over 5500 degrees Celsius, and its incandescence is perceived by us 8 minutes later as sunshine.

The composition of the Sun is far from uniform: about 90 per cent of its mass is located in a sphere of half its total diameter—the central 50 coins. At the outer edge of this region the density is much the same as that of water but it is still hot there: about 4 million degrees. Most of the Sun is hydrogen, but there is a lot of helium in the core, the ash of nuclear fusion, and a spicing of other elements that have been formed as the star burned. In other stars that in the past have exploded, those newly forged elements would have been scattered through the cosmos and some would have settled together as rocky planets and then later, as life evolved and ate the planet, into you and me. We are all, in that wonderful remark that has been said so often before, including here in Chapter 1, made of stardust.

The Sun is middle aged and still a vigorous star. It has been recognizable as a star for over 4.5 billion years and has another 5 billion to go before the remaining hydrogen in its core is exhausted. Its career will follow the path of other largely insignificant stars of a similar mass. The thermonuclear core will spread outwards as the hydrogen there is used up and the heavier ash of nuclear fusion, helium, sinks uselessly into it. The nuclear fusion will spread in a shell around the ever enlarging core and the outer layers will swell up. First, they will engulf Mercury. Then they will engulf Venus. It is not clear whether the bloated Sun, now a Red Giant, will extend to the orbit of the Earth, or whether the Earth's orbit will be so perturbed by the events inside it that it will enlarge and escape engulfment. But it hardly matters. By then, its oceans will have boiled away, its atmosphere will have been blown off the planet, and the Earth will be a cinder in orbit around the Sun. All life, including all the achievements, myths, and fantasies of mankind, if any survive for such a vast length of time, will be gone.

The end days of our currently noble Sun will be quiet. The outer layers that have devoured the inner planets will gradually, after about a million years or so, drift off into space, as though bored with being a star. The inner core will remain fiercely glowing, defiant as a White Dwarf of great density, thousands of times that of water. But its life is limited, for its fuel is depleting and it will gradually cool as it radiates energy into space. Like white hot iron, from White Dwarf the remnant of our star will lapse into Yellow Dwarf, then Orange Dwarf, Red Dwarf, and finally Black Dwarf. This cooling might take a trillion years, but the end of the Sun's days will see it as an ice-encrusted, hugely dense, almost invisible sphere, about the size of the Earth, a single coin, with its light extinguished.

Well before 10 billion years of solar life have passed the human race—or whatever it has evolved into, or whatever has taken its place either on account of the instability of its DNA, its meddling with its own DNA, or its own stupidity—will have to have absconded from its ancestral home and might have leap-frogged around the planets in the search for hospitable environments either for itself or for some avatar-like electronic embodiment of its spirit. There is, however, no escape, for there will be, inexorably, end times. The stars will go out. There will be new generations of stars that follow our own, but they too will go the way of the Sun. Gradually all the stars will go out, everywhere, and none will be born to take their place.

It will be lonelier than even that, on two grounds. First, as the universe expands, so the galaxies move apart. There will come a stage when space has expanded so much that the light of stars in other galaxies will never reach us. We, or what remains of us, shall see the stars in our own galaxy, but gone will be the light from others.

Then the worst will happen. Matter will probably decompose into radiation. I say 'probably' because the long-term future of matter is currently uncertain, but the indications are that all matter will be gone in about 300 000 000 000 000 000

000 000 000 000 000 years, or so. That is indeed unworryingly decently far distant, but that is not the point: science is helping us to see into the infinite future, and finding the absence of everything.

Even the radiation into which the matter—your atoms and mine—has decayed will have gone. As the universe expands, it stretches the waves of radiation, and their wavelength increases. Because we currently think that the universe will expand for ever, and apparently (if our current observations extrapolate into the future), that that expansion is accelerating, then before too long (whatever that might mean), all radiation will have been stretched out flat. Science has revealed God's glorious plan for the universe: it is to go from absolutely nothing to empty dead flat spacetime.

I have to admit, though, that the end days might be different: we really don't know enough yet to be certain about such great lengths of time. Our current experience is with a mere 13.7 billion years, all of it in the past. It could be that the laws of nature are evolving on a time scale of trillions of years, and that the expansion of the universe will not accelerate or even continue forever. It could be that the handful of dimensions that are currently supposed by some to exist but to be tightly coiled and currently invisible to us will uncoil on a timescale of a hundred trillion years, with consequences at present beyond imagination. It could be that dark matter slowly acquires interaction other than gravitation and begins to impinge on our lives. It could be that a defect in spacetime will rip it apart and put us out of our misery as spacetime returns in an instant to Nothing. We simply don't know.

What we do know is that the Sun will run its course, and abandon us, for in its inanimate way it owes us nothing and there is almost certainly nothing we can do to ameliorate its inevitable old age or prolong its vitality. We shall have gone the journey of all purposeless stardust, driven unwittingly by chaos, gloriously but aimlessly evolved into sentience, born unchoosingly into the world, unwillingly taken from it, and inescapably returned to nothing. Such is life.

EPILOGUE

We could simply lie back and think of the creation. Our forefathers could do no more. But careful scientific investigation has displaced wondering and poetic myth by something akin to actual mechanism, and we can see how the entire universe has developed from its initial egg. How that egg was laid is still a mystery, but that initial region of ignorance should not deflect us from appreciating the broad sweep of history of the universe that science has revealed, the cosmic unfolding that has brought us from that extraordinarily pregnant moment to the present day with an understanding that enables us to predict the general features of our future. A part of the extraordinary potency of science is its ability to take measurements on a tiny sample of the universe during a few instants of existence in a nearly microscopic region we call a laboratory over a time sometimes measured in seconds or even less and extrapolate with considerable confidence to the universe as a whole, both in space and time. There are no doubt dangers in doing so, but science believes itself alert to them and seeks to confirm its confidence by measurements on ever greater scales of space and time.

We could simply lie back and think of the biosphere. Our forefathers could do no more. But careful scientific investigation has displaced our total ignorance about the origin of organisms, an ignorance that inspired delightful myths in almost every culture, by a single principle that though difficult to apply seems to have the potency to account for all living species past and present. There remains ignorance about the original egg: how did the inorganic first stumble into becoming the portentous organic. But even here science has not been stumped: it is alive with its ideas but does not yet have sufficient evidence to identify which of them, if any, is correct. What happily happened here can happen anywhere with similarly benign and stable conditions, and it is not an unreasonable supposition that the universe teems with life that has stumbled into being and has evolved by natural selection. This is extrapolation of another kind, and extraordinarily difficult to test, for so great are the times for information to travel from likely candidate worlds and certainly from galaxy to galaxy that what was once alive might well have burned out or has yet to come into being. We might never know: we too might be burned out before the bottle washes up on our shore.

We could simply lie back and think of reproduction, the centrepiece of life. Our forefathers could do no more. But careful scientific investigation has displaced our complete ignorance of what goes on between conception and the risky process of parturition. We saw two aspects of what science has illuminated in the overall chain of events, the cellular mechanism of sexual reproduction and the molecular foundation of inheritance. We should not lose sight of the alliance of simplicity and complexity that nature has adopted to ensure the effective immortality of an organism by periodic replication of the vessel carrying its genes. The simplicity lies in the encoding of the information for the propagation of good approximations to self. Whereas English literature uses over two dozen symbols to express itself, the whole of the biosphere is encoded in just four. The price that nature has to pay for this simplicity is the complexity of mining the DNA and converting information into organism. Both the simplicity and the complexity have been unravelled by the fingers and brains of imaginative scientists and we now think we know the full story of what takes place in pregnancy.

We could simply lie back and think fearfully of death, the unavoidable sequel to life. Our forefathers could do no more, and to ameliorate its prospect built myth upon myth. But careful scientific investigation has revealed what lies in store for all of us, like it or not. There is no comfort here, but science is not about false comfort, it is about truth. Truth is a part of the spirituality of my kind of positive disbelief, for it gives us confidence to employ what talents we have, and if not talents then opportunities. It urges us to use and not to waste the time while we are present on Earth, not unthinkingly to seize the day but judiciously enjoying our brief interlude of consciousness, the flicker of daylight between the dark of the cradle and the grave, for some to comprehend, for others simply to indulge. Science has contributed gloriously to death by furnishing methods to keep it at bay or to ameliorate the painful sharpness of its teeth when finally it visits. Medicine is one of science's great contributions to human well-being, with health arguably more important than education, for it is perhaps better to be alive and ignorant than dead but informed.

We could simply lie back and think hopefully of what follows death. Our forefathers could do no more, and in the absence of any evidence built engaging myths about the variety of come-uppances and rewards that lie in store. Here science proceeds in two strands. It is reasonably secure in its vision when it stands on the edge of the future and peers into the forward abysm of time and sees the end of the world. We can be confident about the middle-term future history of the solar system, which will see the end of all our material, intellectual, and artistic achievements. The very long term future is less certain as it might be the case that unknown principles will come into play and protect us somehow from the current expectation that all there will be is dead flat spacetime. For those secure in the belief that there is life, of some kind, to come science might on this question seem to over-reach itself and bumble in where it is incompetent to tread. Thus Soul slips from its grasp and science being grounded in the pre-afterlife cannot illuminate the afterlife itself. Quite frankly, it is hard to credit that the extraordinary property of complexly interacting organized matter we call consciousness can survive the decomposition of that matter. Science, especially through psychology, shines its brilliant light on the afterlife and instead of illuminating it causes it to shrivel and die, revealing its core: anxiety.

My own faith, my scientific faith, is that there is nothing that the scientific method cannot illuminate and elucidate. Its revelations and insights add immeasurably to the pleasure of being alive. My faith respects the powerful ability of the collective human intelligence, which initially groped for understanding through myth but now gives us the capacity to comprehend and, optimistically, given time and given cooperation between brains, will do so without limit. The scientific method is a distillation of common sense in alliance with honesty, and its discoveries illuminate the world. Unlike myth-making, which is entertainment in alliance with the desperation of sought but thwarted understanding, that illumination is the sound and firm foundation for the joy of true comprehension.

INDEX

INDEX